Intelligent Transportation and Evacuation Planning

Arab Naser • Ali K. Kamrani

Intelligent Transportation and Evacuation Planning

A Modeling-Based Approach

 Springer

Arab Naser
Design and Free Form Fabrication
Laboratory
Industrial Engineering Department
University of Houston
Houston, TX, USA

Ali K. Kamrani
Design and Free Form Fabrication
Laboratory
Industrial Engineering Department
University of Houston
Houston, TX, USA

Fatimah Alnijris's Research Chair
for Advanced Manufacturing Technology
Industrial Engineering Department
King Saud University
Riyadh, Saudi Arabia

ISBN 978-1-4899-7334-4   ISBN 978-1-4614-2143-6 (eBook)
DOI 10.1007/978-1-4614-2143-6
Springer New York Dordrecht Heidelberg London

Printed on acid-free paper

Springer is part of Springer Science+Business Media (www.springer.com)

To my mother, wife and children
                    Arab Naser, Ph.D.

and

To my boys, Arshya & Ariya
                    Ali K. Kamrani, Ph.D., P.E.

"Honesty is the best policy. If I lose mine honor, I lose myself"
                    William Shakespeare (1564–1616)

Preface

For most human beings, facing threats is inevitable. While in some cases a threat can be treated or eliminated, the safest course of action in facing threats is often to evacuate the danger zone. This is especially demonstrated by natural threats, such as earthquakes, hurricanes, and fires, where the force of the threat is much greater than any possible human defense. Evacuation of the danger zone can vary in scale from individual movement to hundreds of thousands of people evacuating urban cities. Urban evacuation can be viewed as the process in which evacuees are moved from danger areas to safe zones utilizing transportation resources. This massive movement of population typically exceeds normal demands on transportation resources and thus requires careful planning and optimization. The greatest risk of a poor evacuation plan is that people may lose their lives if they are not given the chance to evacuate on time. Another less severe consequence is a massive number of people trapped for hours on the road.

Recently, evacuation planning and modeling have attracted interests among researchers as well as government officials. This is probably due to the fact that south eastern states of USA are under increasing threat of hurricane attacks every year. The catastrophic loss of human lives after hurricane Katrina, and the massive evacuation "nightmare" before hurricane Rita in 2005 have also initiated several research efforts for the development of evacuation planning strategies and techniques.

This book presents the state-of-art models and methodologies for evacuation planning. These models integrate various scientific disciplines, such as operations research, simulation, traffic engineering, and systems engineering. Simulation and operations research focus on analytical (mathematical) models and methodologies; thus, the main goal of these models is to find the best routes for evacuation. On the other hand, systems engineering applies the planning process at the top level or design phase, while considering all the following stages. Traffic engineering and intelligent transportation research provides understanding of real life characteristic and attributes for the evacuation process.

Intelligent transportation systems (ITS) are complex systems that apply communications and information systems technology in transportation infrastructure to improve the efficiency, safety, and security of the transportation system. The Research and Innovative Technology Administration of the Department of Transportation (U.S. DOT) lists several research initiatives to accomplish effective ITS. ITS improves safety and mobility of transportation systems and increases productivity of human beings by integrating technologies known for communications, vehicles, and infrastructures used in transportation. A broad range of wireless communications is involved in ITS. In late 1980s, importance of addressing life cycle operations and maintenance as part of the systems development was not known. To address these challenges, systems engineering was introduced to the ITS community. ITS include a vast array of possibilities, and to develop effective and reliable initiatives it is necessary to adopt certain proven methodologies as that presented by systems engineering theories.

Systems engineering is a discipline which deals with the development of both the methodologies and required technologies for design and engineering of complex systems. System is an integrated set of components such as equipments, information, materials, human resources, building, software, etc. which work collectively toward producing a desired outcome based on a set of requirements and specifications. For every system there will be multiple subsystems. Systems can be simply classified into static and dynamic. Many are a combination of these two categories. An example of a dynamic system is transportation systems. Usually large and complex systems are difficult to build and maintain. Systems engineering is an effective process which deals with the overall life cycle design of any system. It manages the projects by using tools and methods stage-by-stage from conception, design, manufacturing, operation, and to disposal.

Mathematical models have been developed for evacuation planning. These models range between network optimization and simulation models. Network optimization models aim at finding "*the best*" routes for evacuation. In evacuation planning, the ultimate goal is to clear and move all evacuees to safety in the minimum amount of time. In pursuing this goal these models have to simplify model complexity by adding assumption that may seem contradictory to real life observations. One of these observations is the relation between roads travel time and the number of evacuees on the road. Most network optimization models assume constant travel time whereas realistically travel time depends on traffic density. Simulation models on the other hand are unable to find the optimum routes. Instead several alternative routes and scenarios are evaluated to assist in the planning process. However, the simple structure of a simulation routine facilitates capturing complex system attributes, such as load-dependent travel time.

An integrated methodology that utilizes the advantages of both models is presented in this book. In this methodology, the goal is to find the best routes that minimize total evacuation time while allowing travel time to be a function of roads density. This book also presents state-of-the-arts models and methodologies for evacuation planning based on the systems engineering concepts.

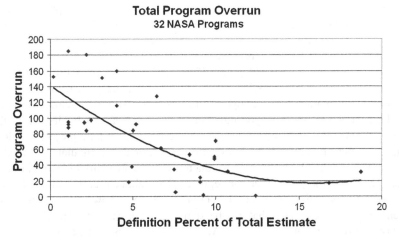

Fig. 1 Cost of project schedule overrun (Understanding the Value of Systems Engineering; Eric Honour; INCOSE 2004)

Chapter 1 presents an introduction to system engineering concept. Systems engineering (SE) is a term that was coined in the 1950s then associated with the process of developing large-scale defense systems. Per INCOSE, systems engineering is: *An interdisciplinary approach and means to enable the realization of successful systems. It focuses on defining customer needs and required functionality early in the development cycle, documenting requirements, then proceeding with design synthesis and system validation while considering the complete problem. Systems engineering integrates all the disciplines and specialty groups into a team effort forming a structured development process that proceeds from concept to production to operation. Systems engineering considers both the business and the technical needs of all customers with the goal of providing a quality product that meets the user needs.*

This definition, while very broad, includes all aspects of a project, from development to disposal. It includes various technical and project management activities as well as design requirements analysis and risk management. In a more general sense, systems engineering is a systematic approach to providing a deliverable with minimal correction costs via detailed and synergistic planning with the development of tools that directly support project management. As shown in Fig. 1[1], systems engineering effort has a direct relationship with the project cost. Furthermore, by mitigating and correcting defects at an early stage via the implementation of the opinions of all parties involved in the project, project costs are kept within the planned range. This extensive planning and coordination effort is a critical aspect of systems engineering.

---

[1]Understanding the Value of Systems Engineering; Eric Honour; INCOSE 2004.

Chapter 2 discusses the application of systems engineering in planning and developing of ITS. The systems engineering process model that is commonly emerging in developing ITS projects is the "V" model and has hence become the standard model per the DOT and FHWA.

Since its creation in the 1980s, the "V" model has taken many different forms and refinements. Namely, the wings of regional architecture(s), feasibility studies, operations and maintenance, changes and upgrades, and retirement/replacement have been added by the transportation sector as an adaptation for its use in the systems engineering of ITS projects. These additions show how project development fits within the entire ITS project life-cycle. For this particular "V" model, from left to right the "V" starts with parameters that involve initial identification of needs. These parameters include the regional ITS architecture, feasibility studies, and concept exploration. The central section of the "V" shows parameters of the systems engineering process that involve defining the project, implementing work, and lastly verifying work completed. At the far right, close-out processes include handing over the system to operations and maintenance, and from there maintaining, upgrading, and/or ultimately replacing the system. As with any infrastructure project, it is important to consider the entire life-cycle. Hence, the wings of the "V" are a key addition that brings the systems engineering of ITS projects full circle. It is clear, by moving from left to right, that the systems engineering approach starts relatively basic and generalized, then moves to highly detailed technical processes, and finally back to full spectrum validation. Further, the "V" is arranged such that quality is assured via verification and validation by linking the left and right sides of the "V" together. To properly achieve detailed design, the larger system is broken down into subsystems, and those subsystems are then decomposed into individual components. As the system is broken down into smaller and smaller pieces, the details of the design and the requirements of the individual components are clearly defined.

Thus, as the systems are broken down, a series of traceable and interlinking documents are generated that set standards for testing, verification, and validation at the system, subsystem, and component levels. These specification documents, while assuring quality, also promote project management. Another crucial component of the systems engineering approach to designing ITS projects is that by decomposing the overall system into smaller and more detailed systems, there is an inherent relationship established between all of the individual stages of the systems engineering process as well as the individual components of the system and subsystems. By documenting and establishing this interrelationship, called *traceability*, project and system requirements are developed intrinsically as part of the design. This is a critical concept of systems engineering that allows the designer to be certain that the system is delivered with the initial needs established at the beginning of the project, and throughout the design stages of the systems engineering process. For example, traceability may include relating a requirement to the subsystem or components that will implement a requirement in the overall system. This traceability ensures that a system or project need will be met through a smaller subsystem requirement.

Chapter 3 outlines the evacuation planning process and introduces the analytical representation of the evacuation models. Evacuation planning and modeling have increasingly attracted interests among researchers as well as government officials. These research efforts use the latest achievements in a number of various disciplines, such as operations research, simulation, and transportation engineering. Evacuation planning involves an iterative process to identify the best routes and to estimate the time required to evacuate the areas at risk. Methods typically used for evacuation planning includes *analytical models* to express the route choice and traffic propagation using mathematical equations and *simulation-based models* to describe the traffic conditions and pattern based on a set of rules. A state-of-the-art review for recent analytical and simulation models and methodologies for evacuation planning is also presented in this chapter.

Chapter 4 presents the proposed integrated methodology for evacuation planning. In this methodology, an optimization algorithm is integrated into a traffic simulation routine in order to determine the best routes for evacuation. Simulation models can predict travel time as a function of flow but they act merely as a tool for evaluating different scenarios and devising recommendations. The main drawback in simulation models is that they do not have the capability of identifying the "*best*" routes (optimization) and therefore a model that integrates an optimization and a simulation routine into one algorithm is proposed and used for evacuation route planning. In the proposed approach, first a traffic simulation model is developed that can propagate a flow along the arcs and predict travel time as a function of arcs load. Then an optimization model is developed to minimize total evacuation time. It is based on network optimization formulations and principles. The next step in the methodology will integrate both models. The main idea is to discretisize the time space and to solve both routines at each time step.

Finally, several sample cases using the proposed methodology are presented in Chap. 5. The purpose of this chapter is to test the premise of the presented methodology. To achieve this goal, a traffic simulator is developed that evaluates the optimum solution obtained by a constant travel time algorithm. The application considered is evacuation planning, and therefore, the selected constant travel time algorithm is for the quickest transshipment problem.

Houston, TX, USA

Arab Naser, Ph.D.
Ali K. Kamrani, Ph.D., P.E.

Acknowledgments

We would like to thank our reviewers for assisting us to improve the content and the quality of our book. We would also like to thank Springer (USA) for giving us the opportunity to fulfill this project.

Acknowledgements

Contents

Chapter 1
Introduction to Systems Engineering

Abstract Systems engineering (SE) is a term associated with the process of developing large-scale complex systems. By mitigating risks and correcting defects during the early stage via the application of the systems engineering process, experience has shown that project costs and risks are significantly reduced. This chapter outlines an introduction to the concept of systems engineering and its processes. It also lists major advantages of its implementation as part of the development of any complex systems.

1.1 Introduction

Systems are an assemblage of objects united by some form of interaction and/or interdependence to achieve a set goal. They are a combination of many resources, such as humans, materials, equipment, hardware, software, facilities, data memory, etc. A system requires a large amount of data, expert personnel, equipment, facilities, and proper management for effective operation. The scope of a system is illustrated in Fig. 1.1. A system is based on hierarchy as shown above. This allows for better definition, decomposition, and the final integration of the system. An example of space transportation systems hierarchy is illustrated in Fig. 1.2. Systems engineering is a discipline that deals with the development of both the methodologies and the required technologies for the designing and engineering of complex systems.

A system is an integrated set of components, such as equipment, information, materials, human resources, building, software, etc., which work collectively toward producing a desired outcome based on a set of requirements and specifications. For every system, there will be multiple subsystems.

Systems can be simply classified into static and dynamic. Many are the combinations of these two categories. An example of a dynamic system is the transportation systems. Buildings, bridges, and roads are considered static systems. Usually, large and complex systems are difficult to build and maintain; therefore,

A. Naser and A.K. Kamrani, *Intelligent Transportation and Evacuation Planning: A Modeling-Based Approach*, DOI 10.1007/978-1-4614-2143-6_1,

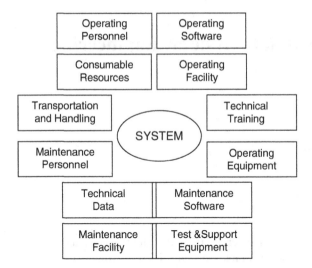

Fig. 1.1 Scope of a system

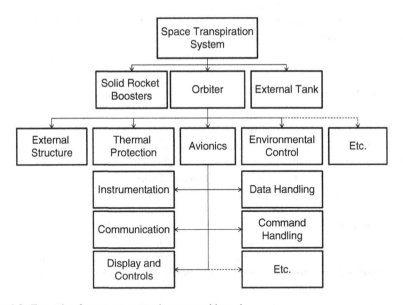

Fig. 1.2 Example of space transportation system hierarchy

the systems engineering process is required for the design and development of such systems. Systems engineering is an effective process, which deals with the overall life cycle design of any system. It manages the projects by using tools and methods stage-by-stage from conception, to design, to production, to operation, and finally, to disposal. It is a technical management technique. Systems engineering is an integrated discipline of technology, science, organization, and management. It interprets the needs of the customer and integrates them in the system to be produced while considering the overall life cycle. There are four major elements within the systems engineering process that must be carefully processed when applying the systems engineering models and techniques. These four elements are described below.

1.2 Manage Uncertainty and Risk in the System

A certain level of uncertainty always exists in any design of complex systems. Uncertainty in design will increase as the complexity of the system increases. Uncertainty could be due to the environment that the system operates in, lack of clear understanding of the requirements, poor communication, complexity, etc. To better manage the uncertainty issues with the design of the systems, the systems engineer must identify the problems that can lead to system's failure. The systems engineer must implement the necessary steps to first analyze the level of uncertainty and then measure the risk level(s) if something goes wrong. Finally, the systems engineer must implement steps to mitigate the uncertainty. There are several methods available that can be used for handling uncertainty. These may include failure mode effect analysis (FMEA), simulation modeling, and prototyping.

1.3 System Quality

Systems must perform as defined by the stakeholders. Stakeholders will define the required performance and functionality of the system, including customers (e.g., operation), corporations (e.g., profit), and the government (e.g., environment). Many factors will impact the quality of the designed system. These factors may include the level of complexity, reliability, availability, and maintainability. Proper decomposition will contribute to the better design of complex systems. Reliability is the probability that a system will operate its intended function without failure for a period of time under specific operating conditions. Availability is the probability that a system is available with full capability. Maintainability is measured based on ease to maintain, amount of time, and cost it will take to get the system operational.

1.4 Project Management

A good management strategy will always contribute to the successful implementation of complex systems. The basic steps required for good management are proper planning, well-scheduled tracking, and timely control. Gantt chart is one of the key tools used for monitoring, scheduling, and controlling activities. It is also used for proper resourcing and cost management. The critical path method (CPM), work breakdown structure (WBS), and resource utilization charts are a few examples of the methods that are available for management of projects.

1.5 Systems Engineer Responsibility

Systems engineers must have an overall knowledge and understanding of the tools and methods used for the systems engineering process. The systems engineer must always see the big picture for the project under consideration. One of the most important tasks of the systems engineer is to understand interactions and interdependencies and ways to manage them. Systems engineers must also have good communication skills and the ability to work within a multidiscipline group of engineers and scientists. Systems engineers must have the ability to communicate with teams from different backgrounds and be able to communicate the views to all the groups accurately. The systems engineer must be flexible to manage changes through the system design and development. These are the key attributes of a good systems engineer.

1.6 The Life Cycle of Systems Engineering

The stages of systems engineering are the design life cycle for the project. These steps are briefly described in the following sections.

- *Conception*: In this stage, the systems engineer must outline the overall system capabilities, systems mission or purpose, identify system boundaries, and to finally identify interactions of the system with its environment and with other systems. The concept of operations describes the layout of the system and how it should operate.
- *Requirements analysis*: Requirement analysis will address the needs of the stakeholders. At this stage, the systems engineer must analyze both the high level and detailed level of requirements until it is clear to all stockholders of what customer needs are and how it should be addressed. After this stage which is considered to be most critical, the requirements are transformed into functions and systems specifications.
- *Design*: In this phase of the process, the team will evaluate the alternative design generated using trade-off analysis and techniques. Preliminary design review process will be done for each level whether the design is compatible with defined requirements and what are the initial risk levels and required mitigation. During the critical design review process, the final design of the system is selected.

- *Implementation*: This phase deals with the actual fabrication of the subsystems. During this stage, the system is being implemented based on the defined specifications. It will ensure that the required quality at each phase of implantation is met.
- *Testing*: The objective of the testing is to make sure that the system is working as per defined conditions and producing the intended performance. There are three levels of testing during the design life cycle. These are component or unit testing, subsystem testing, and integration testing.
- *Integration*: During integration the system is assembled. This is a critical stage of development since the individual behavior of the components and the subsystems now have to perform in an integrated mode and provide the required functionality and performance. The integration testing at this stage will assure the final quality of the assembled system.
- *System acceptance*: After the completion of the system, the final product must be accepted by the customer. For this purpose and system acceptance, a plan is required. In this plan, a comprehensive set of requirements is provided for accepting the final system. This plan must be reviewed and accepted by both the customer ordering the system and the group that is developing the system.
- *Operation and maintenance*: After the delivery, the system must be maintained. There are typically two categories of maintenance, excluding the corrective one. These are (1) preventive maintenance and (2) predictive maintenance. The systematic inspection, detection, and correction of failures either prior to their occurrence or before they develop into a major defects is preventive maintenance. Predictive maintenance techniques allow for determining the condition of in-service equipment in order to predict when maintenance should be performed. This approach offers cost savings over routine or time-based preventive maintenance.

1.7   Elements of Systems Engineering

Per the systems engineering capability maturity model (SE-CMM), systems engineering is based on process and capability dimensions. Process dimensions describe the steps that an organization implements needed for good practice of systems engineering. The capability dimension focuses on the capability or how well an organization performs the SE steps. Basic systems engineering processes are divided into three processes. These are organizational, project, and engineering processes. These processes are further divided into the following areas.

Organization processes: This process provides the basic infrastructure that supports the organization. The processes include the following:

- *Coordinating with suppliers* includes processes such as make or buy decisions, choosing the suppliers by evaluating their capability, steps for understanding of

what is needed, timely audits and inspection of supplier's products quality, communication, etc.
- *Managing systems engineering processes* provides the required infrastructure and resources needed for the organization to manage complex projects. The processes include techniques such as optimization, simulation, and other analytical tools required for managing organizational processed.

Project processes: There are five elements within the project process area. These elements are required for the success of the project. These are as follows:

- *Quality control* means that the system must be operating according to the customer's requirements. Quality control must be implemented in all stages to ensure that the system is being developed per defined specifications. Quality can be maintained in two ways. First, quality must be incorporated into all of the stages of the process, and next is to test the final product. Several tools are available for quality control, such as acceptance sampling techniques: Pareto charts, six-sigma methodology, statistical quality control, inspections, etc.
- *Configuration management* is the relative arrangement of the elements. It creates a base line for every system and manages the system itself and also any changes that may occur during the development process. It focuses on establishing and maintaining the consistency of a system or product's performance and its functional and physical attributes with its requirements, design, and operational information throughout its life An advantage of a configuration management application is that the entire collection of systems can be reviewed to make sure any changes made to one system do not adversely affect any of the other systems.
- *Manage risk*: Risk management is an important aspect of any design. The systems engineer must have the necessary tools and experience to estimate the impact of the risk and the cost required for mitigating the risk. To reduce the risk in case of technical failures, there are several techniques, such as simulation and prototyping to assess the performance impact of the system prior to its implementation.
- *Monitor and control technical effort*: Project management techniques are used for planning, monitoring, and controlling projects. Tools such as the Gantt chart and CPM are used to determine the duration of the project. Proper steps must be taken to use these tools properly and to identify all activities of the project to avoid any delay in the project.
- *Plan technical effort*: Planning is an important aspect of any project. Tools such as a WBS can be used to breakdown the project into smaller elements for effective planning. WBS is a structured approach in which the work is divided into work elements. For each tasks and work elements, the duration, responsibilities, requirements, and resources are defined.

Engineering processes: Steps within engineering processes play a critical role in implementing a successful project.

- *Integrating disciplines*: Most projects require resources from different disciplines. A multidisciplinary team is required to implement a successful system.

Members from different disciplines such as mechanical engineering, electrical engineering, chemical engineering, industrial engineering, management, business, and software/hardware. How the teams are managed depends on different organizational paradigms. Examples of these paradigms are functional project-oriented, project-coordinated, or matrix organizations.

- *Understanding customer needs*: Understanding customer need and its interpretation into the requirement is the "key factor" in the successful implementation of any project. There are several cases where after the completion of the system, the system did not deliver the required performance and functionality due to lack of understanding the customers' needs. Typically, a concept of operations document is used to prepare a comprehensive list of what the requirements are based on the operational needs. At this stage, the technical performance measures are also determined.

- *Requirements allocation* takes place once there is a clear understanding of what the customer needs and requirements are, they are further transformed in terms of the functions and specific operations of the system. The high level requirements are decomposed into more levels of detail for ease of allocation and implantation. This decomposition should allow the systems engineer to identify and allocate subsystems and components within the system.

- *Traceability matrix* is a closed-loop systems engineering process. This matrix is used to make sure that all requirements are allocated to the elements within the system. If a single requirement is not properly addressed, it will make a big difference to the final output of the process and could lead to a dysfunctional design.

- *Systems architecture decomposition and integration*, usually a top-down approach, is used to decompose the architecture and the overall framework of the system. Each system is divided into a number of subsystems, and each subsystem can also be further divided. This is an iterative process until all the details regarding the system and its components are clearly defined. This also allows the SE to identify the interaction and interrelation between the elements within the system. These may include physical connections, logical correlations, and information flow. All subsystems including components are then integrated into one overall system to achieve the final design and required functionality.

- *Verification and validation* are processes that make sure the developers have developed the right system and it is operating as intended. Acceptance testing is done at the customer's site for the final approval of the system by the user.

1.8 Systems Engineering Paradigms

Past experience has shown that lack of planning and clear identification of objectives has been the major problem associated with the design and development of complex systems. This approach has resulted in systems that lack performance and designs that ultimately fail. Traditionally, systems have been developed based on "Deliver now and Fix Later." [1]. Process has suffered from a lack of clear planning, which results in failure and a high cost of design modifications. In this scenario, requirements at the systems level were kept general in order to reduce complexity to allow for new technological integration. This has routinely evolved into last minute modifications that impacted the schedule and cost. Decisions made at the early stages of the development life cycle will significantly impact the overall life cycle, including cost and system's effectiveness. Therefore, there is a need for a disciplined approach for integrated design and the development of new systems. In this case, all aspects of the development are considered early in the process and used for continuous improvement. Per Department of Defense (DOD) [2], systems engineering is the effective application of scientific and engineering efforts to do the following:

1. Transform an operational need into a defined system configuration through the top-down iterative process of requirement analysis, functional analysis and allocation, synthesis, design optimization, test, evaluation, and validation.
2. Integrate related technical parameters and ensure the compatibility of all physical, functional, and program interfaces in a manner that optimizes the total definition and design.
3. Integrate reliability, maintainability, usability, safety, serviceability, disposability to meet cost, schedule, and technical performance objectives.

Systems engineering is also considered a process for managing technology. The systems engineering process has evolved in seven different paradigms [3]. A summary of selected processes are discussed below.

Build–test–fix: This method consists of three basic steps: fabricate a design, test the system, and then operate. This method is typically used for in-house software development where the customer is also the developer. It is considered to be a simple but effective approach although it lacks the requirements analysis phase that makes it not suitable for any complex systems design.

Staircase: The staircase method allows for better management and control of systems development. This method is considered to be a systematic flow through the SE process. It is well-suited for the developments of existing systems variants. In this case, already developed requirements are modified. The export version of a military aircraft is an example of this concept: Requirements–Specification–Design–Fabrication–Testing–Acceptance–Operate area steps in the staircase SE cycle. An improvement to this model has incorporated feedbacks into all the phases. This model is known as Staircase with feedback.

Waterfall: This method improves the staircase method by adding feedback loops between successive stages as shown in Fig. 1.3. Through these feedbacks, each stage is capable of gaining knowledge from the subsequent stages. The success of this method depends on understanding and processing revisions through feedback analysis.

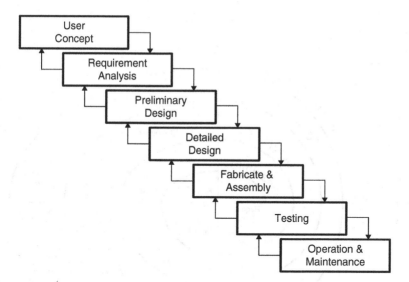

Fig. 1.3 Waterfall systems engineering process model

Early prototype: The early prototype process is an extension to the staircase with feedback cycle. This method is used when it is difficult for the customer to identify requirements, but could recognize them through some model or prototype representation. The advantage of this method is due to the direct interaction between stakeholders. Some of the difficulties with this approach are (1) the initial prototype could discourage the customer, (2) it suggests an unattainable goals, and (3) prototype design becomes the main objective rather than the actual customer's need.

Spiral: The spiral method (Fig. 1.4) is an extension to the early prototype concept. The primary advantage of the spiral method is the detailed development of requirements, specifications, and designs. The significant challenge for the spiral method is managing the prototype transitions. Some of the advantages are the following:

- Risk-driven sequential phases with user involvement.
- Considers highest risks issue first (requires understanding; technical feasibility, and system operations).
- Cycles of risk-driven phases, spiral around and end with a final waterfall wrap.

Integrated: The integrated or concurrent development method consists of cross-functional teams with members from all of the functional areas working closely together, sharing details of their portion of the design as it progresses, and developing all aspects of the system simultaneously. The result is overlapping and managing the overall life cycle. Concurrent engineering (CE) is defined as the systematic and integrated approach to the systems life cycle design. CE is also known as the design for life cycle model. Concurrent engineering is the implementation of parallel designs by cross-functional teams including suppliers. Without empowered team members and the free flow of communications, this method will not function.

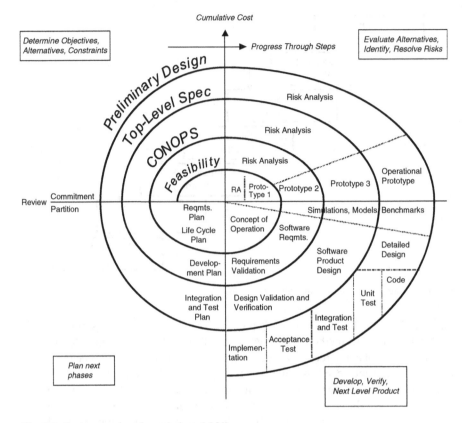

Fig. 1.4 Systems engineering spiral model [4]

"V" model: The "V" model is a graphical representation of the systems development life cycle. It illustrates the necessary processes to be undertaken within the systems development life cycle to achieve a successful systems project. It also shows the holistic nature of the process by depicting the direct relationships between the decomposition and definition stages and the integration and recognition phase.

This concept is based on the iterative and parallel processes on the left-hand side that will manage the verification functions on the right-hand side. Verification is completed in a serial fashion, resulting in minimal rework. This method is cost effective and improves the success of the project. It also provides the necessary infrastructure for alternative design analysis and selection. It is believed that by using this approach the probability of designing a reliable and satiable system is high. The "V" process (Fig. 1.5) is built on the following layers:

1. *User's perspective*: This is the perspective of the customer or stakeholder that defines a set of requirements and expects a finished product that meets the defined requirements.

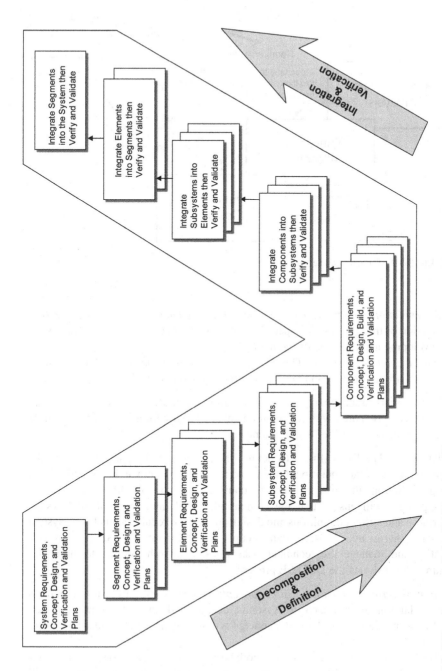

Fig. 1.5 "V" process as applied to NASA projects [5]

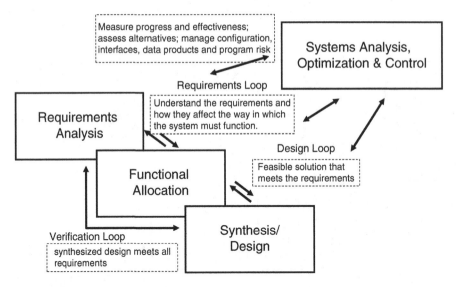

Fig. 1.6 DOD SE process cycle [2]

2. *Systems engineer's perspective*: The architectural details that address the decomposition of the system-level specification into system design. It also includes the development and testing of the final product.

3. *Implementer perspective*: This is the implementation process, which includes specifications, design, and testing, and the component level.

These layers assure that the design performance and feasibility (schedule) are continuously compared with the requirements and allow for analysis of alternative designs against the verified requirements.

DOD model: The DOD (Fig. 1.6) defines the systems engineering process as the transformation of the operational needs and requirements into an integrated system design solution through concurrent consideration of all life cycle needs. This process will ensure that the compatibility and integration of all physical interfaces and system definitions and designs reflect the requirements for all system elements (hardware, software, data, people, etc.). Finally, the SE process will identify and manage the technical risks associated with system development. Figure 1.6 illustrates the DOD SE process.

Simulation-based acquisition (SBA) concept: The SBA concept is the integrated and collaborative approach to systems design and through computer-based modeling and simulation. The SBA concept is based on the DOD acquisition reform initiative.

The SBA concept is based on the collaborative engineering concept and environment. Industrial partners, academia experts, and government agencies will

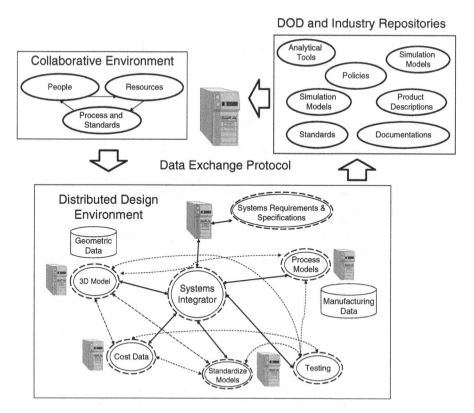

Fig. 1.7 SBA environment

collaborate using technologies, developed methodologies, and resources. This will reduce the development time and cost-associated increased performance and functionality. The principal architectural concepts used for SBA implementation are the following:

1. Collaborative Environment and Reference Systems Architecture
2. Distributed Product Descriptions and Data Exchange Format
3. DOD/Industry Resource Repository

SBA is not a replacement for the systems engineering process [6]. It is a distributed and integrated approach to design, using the systems engineering principles. It is a modeling and simulation (M&S) technique used to support managers during their decision-making process. It must maintain the integrity and security of all shared data, including responsibility and accountability at all levels of proprietary and security. Figure 1.7 illustrates the SBA concept. Table 1.1 summarizes some of the advantages and disadvantages of the sample SE processes [7].

Table 1.1 Advantage and disadvantages of sample SE models

Model	Advantages	Disadvantages
Waterfall	• The oldest and most widely used method • Previous cases are well documented • Accepted and easy to understand by customers • The model is a logical sequence of processes • Clear and defined tasks and their boundaries	• Sequential flow • Requires definition of high level of accuracy details at the early stage • Lack of mechanism for detecting errors until the end of the cycle
Spiral	• Multiple deliveries of systems • Good model for software intensive systems • Focused on identifying and mitigating risks • Risks analysis prior to proceeding to a next phase • Identification of high-risk requirements and evaluated using prototyping • Applied to system components individually and independently • Customer approval before proceeding to the next phase	• Required prototype development and redevelopment • Spiral is based on waterfall development • Overall planning complexity and costing • Deliverables are not well defined • Prototypes mainly used only for risk analysis and not actual development • Intense customer involvement and approval
V	• Integrated approaches • Systematic decomposition analysis of requirements into preliminary and detailed designs • Systematic integration and verification analysis of system components including testing	• Feedback-based process • Integration planning during the design requirements phase • Verification planning during the integration phase

1.9   Addressing Uncertainty and Risks in Systems Engineering

One of the key factors in being able to analyze and measure the degree of success in a systems engineering project is the ability of the system engineer to manage and overcome uncertainty and risks. Some of the factors that cause uncertainties and risks in projects according to the US Department of Transportation are discussed here [8].

Complexity: The more parts a system has, the more complex it becomes. This complexity can contribute to poor integrations and management. Complexity in the system brings uncertainty in two different ways. First, complex systems are difficult to visualize and understand. This could lead to poor interpretation

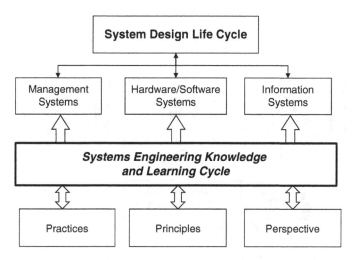

Fig. 1.8 Scope of learning cycle in SE process

requirements and how to address them. This could result in poor design and a dysfunctional system. Proper decomposition of the system must be used to reduce the level of complexity. Second, it would be difficult to define and maintain interactions within the system. Changes in complex systems must be carefully addressed and managed; otherwise, it could lead to undesirable results.

Technological change: Given the rapid change in technical capability, projects always face uncertainty in choosing the right technology. For example, the changes in software technologies occur at a fast pace. Systems engineers and managers must implement a strategy of how to select what technology to use if they are going to implement a system that is based on advanced technology. Information imperfection: This is a major portion of the SE process in information processing and knowledge development. The goal of systems engineering is to determine whether information is missing and if there is a need to improve the existing information. Figure 1.8 illustrates the scope of the SE learning cycle and its impact.

1.10 SE Tools for Addressing Uncertainty and Risks

The US Department of Transportation considers seven tools to address uncertainty and risks [8]:

Project scheduling: There are three basic scheduling and tracking tools. These are Gantt charts, activity networks, and WBSs. An activity network shows tasks and the order in which they are processed. This type of network is usually used to determine the project critical path, which is the sequence of activities that could delay the

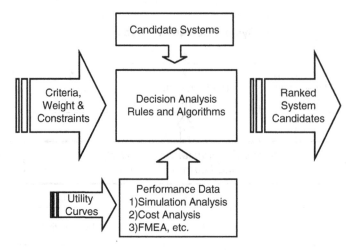

Fig. 1.9 Trade-off studies and analysis

implementation of the project. Project managers must carefully monitor the critical path cycle and timing.

Trade-off studies: Trade-off studies are often conducted as studies where the analyst will assess the vendor's solutions using the technical descriptions provided. The analyst constructs a table and then assesses the technical descriptions of the products against the requirements. An alternative approach for conducting trade-off studies is by testing proposed solutions in a controlled environment that simulates the actual conditions of the system. The implementation of this type of study is not always possible, but it is the most reliable. With the advancement of computer technologies, this approach is becoming more and more applicable. This type of study can also address the technological change. Figure 1.9 illustrate the scope of trade-off studies as proposed by DOD [9].

Reviews and audits: Design reviews are to ensure that the system will meet the defined requirements and needs of the end-use. These reviews are done at different stages of the design life cycle. Examples of these reviews are systems requirement review, systems design review, systems specification review, preliminary design review, and critical design review. The major goal of these reviews is to ensure that the system is understood and designed as defined. It is intended to find issues that are corrected before any implementation.

Technical performance measures: TPMs are established early in the project. TPMs are used to measure the performance of the design and its process. Design modifications are implemented based on the analysis of the TPMs during all stages of design in order to achieve the defined goal. Figure 1.10 illustrates the steps for selecting TPMs [9].

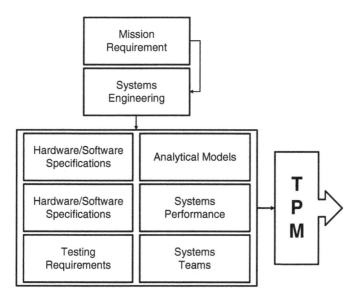

Fig. 1.10 Selection cycle for TPM

Benchmarking: Benchmarking is used on the implemented version of a system as part of the understanding of its operation and features and also used for acceptance testing of the final product. The steps include (1) plan, (2) search, (3) observe, (4) analyze, and (5) adapt.

Prototyping: Prototypes are operational models of the system, but typically in a scale down format. Prototypes are very useful for establishing requirements and better understanding of required performance. For example, prototypes are used to evaluate whether it is feasible to interface elements of a system that have never been linked together. This allows for evaluating the risk impact and reduces uncertainty related to interfacing elements within the system.

Modeling and simulation: Modeling and simulation methods help to examine potential bottlenecks or impediments for the performance of the system. Once the potential problem areas are identified, the designer can modify in order to improve the design of the system. These tools are very effective in reducing uncertainty and improve system performance.

Chapter 2
Systems Engineering and Intelligent Transportation Systems

Abstract Intelligent transportation systems (ITS) are complex systems that apply communications and information systems technology in a transportation infrastructure to improve the efficiency, safety, and security of the transportation system. The systems engineering process model that is commonly emerging in developing ITS projects is the "V" model and has hence become the standard model per the DOT and FHWA. This chapter discusses the fundamental application of systems engineering process in planning and developing of ITS.

2.1 Introduction

Some aspects of systems engineering have been described in the previous section. It is an interdisciplinary field that encompasses various technical expertise to successfully implement a complex project. The International Council on Systems Engineering (INCOSE) defines systems engineering as "An interdisciplinary approach and means to enable the realization of successful systems." It adopts a holistic view that focuses on identifying and itemizing customer needs and required functionality at the initial stage of the development cycle, tabulating the requirements, and then engaging in the design synthesis and system validation while still taking into consideration the complete problem of the system life cycle. Furthermore, INCOSE states that "systems engineering integrates all the disciplines and specialty groups into a team effort forming a structured development process that proceeds from concept to production to operation. Systems engineering considers both the business and the technical needs of all customers with the goal of providing a quality product that meets the user needs."

A. Naser and A.K. Kamrani, *Intelligent Transportation and Evacuation Planning:*
A Modeling-Based Approach, DOI 10.1007/978-1-4614-2143-6_2,
© Springer Science+Business Media, LLC 2012

2.2 Intelligent Transportation Systems

Intelligent transportation systems (ITS) are complex systems that apply communications and information systems technology in a transportation infrastructure to improve the efficiency, safety, and security of the transportation system. The Research and Innovative Technology Administration of the Department of Transportation lists several research initiatives to accomplish effective ITS. Some of these are the following [10]:

- *Vehicle-to-vehicle (V2V) communications for safety*: This involves the use of a dynamic wireless exchange between two or more vehicles. The anonymous exchange of data such as speed, distance, and travel direction enable the vehicles to sense hazardous conditions to mitigate adverse circumstances. An illustration of this is depicted in Fig. 2.3.
- *Vehicle-to-infrastructure (V2I) communications for safety*: This is the wireless exchange of information between the vehicles and the highway infrastructure to avoid car accidents and improve environmental conditions.
- *Real-time data capture and management*: It is the formation of accessibility to real-time and archived multimodal transportation information that is retrieved from vehicles, infrastructure, and mobile devices.
- *Dynamic mobility applications*: This endeavor attempts to identify, develop, and implement applications that enable the vehicles and transportation infrastructure to enhance its present operational configuration.
- *Applications for the environment of real-time information synthesis (AERIS)*: This program involves the use of a real-time vehicle, infrastructure, and transportation data to achieve environment friendly transport systems. AERIS uses a multimodal approach and operates in concert with the vehicle-to-vehicle initiative.
- *Road weather management*: This program involves the use of real-time vehicle and infrastructure data along with data of weather conditions to enhance the decision-making process and forecast capabilities.

ITS will improve the safety and mobility of the transportation system and increases the productivity of human beings by integrating technologies known for communications and vehicles and infrastructures used in transportation. A broad range of wireless communications is involved in ITS. In late 1980s, importance of addressing life cycle operations and maintenance as part of the systems development was not known. To address these challenges, systems engineering was introduced to the ITS community. ITS include a vast array of possibilities, and to develop effective and reliable initiatives it is necessary to adopt certain proven methodologies as that presented by systems engineering theories.

2.3 Systems Engineering Approaches to Designing ITS

There are two types of ITS projects: low risks and high risks. A low-risk ITS project usually has the following characteristics:

- No new software is required.
- Off-the-shelf technologies.
- The requirements of the system are well defined and available.
- The systems operations techniques are well documented.

Figure 2.1 illustrates the sequential SE process for low-risk ITS.

This is considered as a sequential process due to the nature of the design where many of the aspects are adopted from other designed templates. An example of low-risk ITS is to add new closed circuit cameras to a system which currently has cameras. A high-risk ITS project has some or all of the characteristics as follows:

- New software should be used.
- New hardware and traditional technologies should be integrated.
- Interfaces for new or existing systems should be developed.
- Occurrence of changes is continuous.

Examples of ITS project are:

- *Event management*: Sensors and cameras, traffic control, lane management, parking management, information dissemination, enforcement, and special events transportation management
- *Freeway management*: Ramp control and enforcement
- *Road weather management*: Surveillance, monitoring and prediction, information dissemination, traffic control, and response and treatment

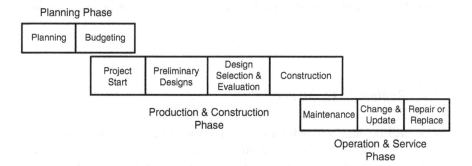

Fig. 2.1 Proposed sequential cycle for ITS

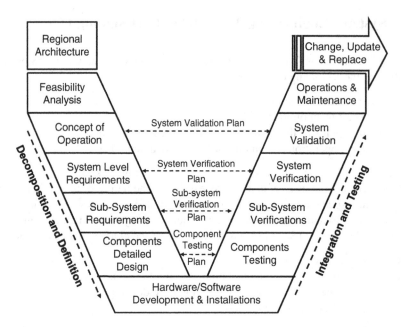

Fig. 2.2 Systems engineering "V" diagram

- *Roadway operations and maintenance*: Asset management and work zone/incident management
- *Travel information*: Pre-trip information, en route information, and tourism and events

Systems engineering process has valuable applications in the design of ITS. Figure 2.2 illustrates the "V" model for high-risk ITS design [10].

The different phases of this process are further discussed in the following sections [10–12].

2.3.1 Regional ITS Structure

The regional ITS architecture is the basis of implementing the systems engineering processes and analyses. It is used to provide regional information of the project, which can be expanded on during the later stages of the project. Figure 2.3 illustrates the context proposed for this phase.

Some key activities used in the initial stage of the ITS architecture are the following:

- Identify regional ITS architecture(s) that are relevant to the project
- Identify the portion of the regional ITS architecture that applies

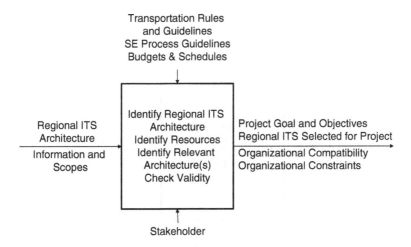

Fig. 2.3 IDEF0 activity diagram representation (partial) for regional ITS analysis

- Verify consistency with the regional ITS architecture and identify any necessary changes to it

  The objectives are the following:

- Define project scope while considering the regional vision
- Identifying areas of opportunities for task integration
- Improve consistency among activities in ITS projects
- Identify efficient incremental implementation strategies
- Integrate between project planning and development

  The sources of inputs (I) and outputs (O) include the following:

- Relevant architecture (I)
- Regional/national resources (I)
- Planning programming products relevant to the project (I)
- Requirements of the proposed system (O)
- List of interfaces shared by the system (O)
- Information to be exchanged by the system (O)
- Project stakeholders, roles, and responsibilities (O)

2.3.2   Feasibility Analysis

The next stage following the "V" process is the feasibility analysis. In this stage, the economic, technical, and political feasibility is evaluated. The benefits and risks are calculated and the major risks are classified. Figures 2.4 and 2.5 illustrate

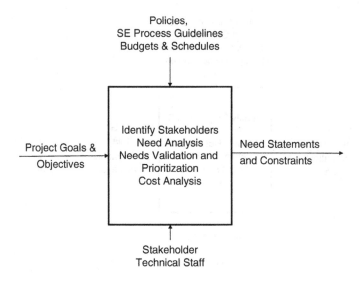

Fig. 2.4 IDEF0 activity diagram representation (partial) for need analysis

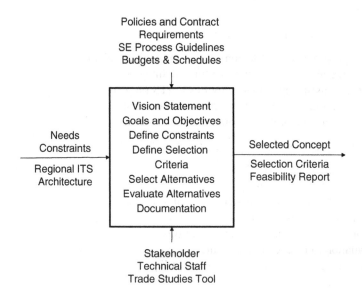

Fig. 2.5 IDEF0 activity diagram representation (partial) for feasibility analysis

the scope of the needs assessment and feasibility analysis. Some of the key activities are the following:

- The definition of the evaluation criteria
- Identification of alternative concepts

- Evaluation of alternative concepts
- Documentation of the results

The objectives are the following:

- Identify cost-effective concept and document
- Criteria and rationale for concept feasibility analysis and selection
- Verifying the proposed project feasibility
- Identify preliminary risks
- Management approval for the project implementations

The sources of inputs (I) and outputs (O) include the following:

- Proposed goals and objectives (I)
- Project purpose and need (I)
- Project scope (I)
- Information regarding the regional ITS architecture (I)
- Feasibility study steps and results (O)
- Proposed alternative concepts (O)
- Proposed business case for the project and selected concept (O)

2.3.3   Concept of Operations (ConOps)

The third stage is the concept of operations (ConOps). In this stage, the project stakeholders reach a mutual understanding of the systems to be developed and implemented. They also discuss how it will be operated and maintained. Typical information included in ConOps is listed below:

- Description of the major phases
- Operational scenarios and timelines data architecture and format
- Operational facilities including integrated logistics support
- Critical events

Figure 2.6 illustrates this scope of this phase of the process.
The key activities in the stage are the following:

- Identification of the stakeholders associated with the system/project
- Define the core group responsible for creating the concept of operations
- Develop the initial ConOps, review it with the broader group of stakeholders, and iterate
- Define stakeholder needs
- Create a systems validation plan

The objectives are listed below:

- High-level identification of user needs
- Identification of system capabilities in terms

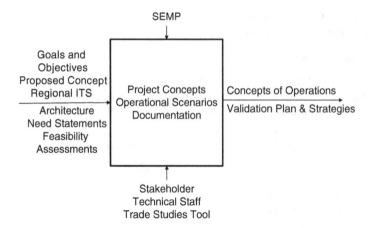

Fig. 2.6 IDEF0 activity diagram representation (partial) for ConOps analysis

- Stakeholder roles and responsibilities for the system
- Understanding developers on the who, what, why, where, and how of the system
- Identification of key performance measures and validation process

  The sources of inputs (I) and outputs (O) include the following:

- Stakeholder roles and responsibilities (I)
- Information lists for other components from the regional ITS architecture (I)
- Recommended concept and feasibility study (I)
- Concept of operations describing the who, what, why, where, and how (O)
- System validation plan and the approach used to validate at point of delivery (O)

2.3.4 System Requirements

In this stage, the stakeholder needs are transformed which allows for the definition of what the system should accomplish and how it will be accomplished. The key activities are listed below:

- Elicit requirements
- Analyze requirements
- Validate requirements
- Manage requirements
- Validate requirements

  Figure 2.7 illustrates the scope of this phase. The objective is to develop a validated set of system requirements that meets stakeholders' needs. The sources of inputs (I) and outputs (O) include the following:

- Concept of operations
- Applicable ITS standards from the regional ITS architecture (I)

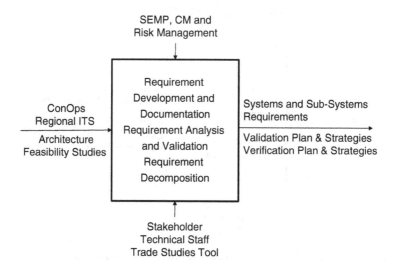

Fig. 2.7 IDEF0 activity diagram representation (partial) for requirement analysis

- Applicable regulations and policies (I)
- Constraints (I)
- System requirements document (O)
- System verification plan (O)
- Traceability matrix (O)
- System acceptance plan (O)

2.3.5   Systems Design

Stage five is the systems design stage. It is generated based on the system requirements, which include a high-level design that defines the overall framework for the system. The key activities are listed below:

- Develop high-level design alternatives
- Evaluate the alternatives
- Analyze and allocate requirements
- Document the interface
- Detailed design

Figures 2.8 and 2.9 illustrate the scope of high-level and component/unit-level designs.

The objectives are listed below:

- Produce a high-level design that meets the system requirements
- To identify and define key interfaces for proper integration

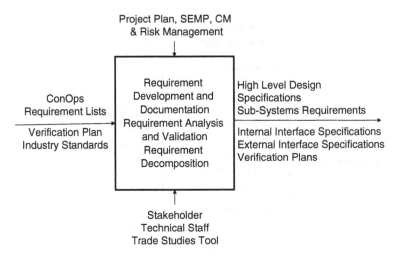

Fig. 2.8 IDEF0 activity diagram representation (partial) for high-level design

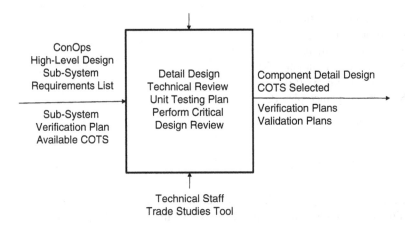

Fig. 2.9 IDEF0 activity diagram representation (partial) for unit-level design

- Develop detailed design specifications
- To select best COTS (commercial-off-the-shelf) products

  The sources of inputs (I) and outputs (O) include:

- Concept of operations (I)
- System requirements (I)
- Off-the-shelf products (I)
- Existing system design templates (I)
- ITS sand industry standards (I)

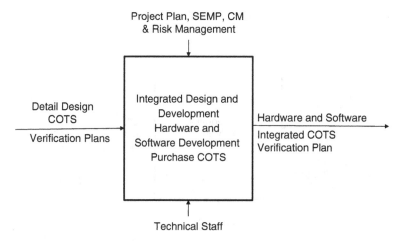

Fig. 2.10 IDEF0 activity diagram representation (partial) for system hardware and software development

- Selected off-the-shelf component (O)
- High-level and unit-level designs (O)
- Hardware/software specifications (O)
- Unit test plans (O)
- Subsystem verification plans (O)
- Acceptance plans (O)

2.3.6 Hardware and Software Design

The next stage in the SE cycle is the software/hardware development and testing. In this stage, hardware and software solutions are created for the components identified in the system design. The objective of this step is to develop and purchase necessary hardware and software that meets the design specifications. Figure 2.10 illustrates the scope of this phase. Key activities are listed below:

- Plan software/hardware development
- Purchase off-the-shelf products
- Develop software and hardware
- Develop supporting products
- Perform unit/device testing

   The sources of inputs (I) and outputs (O) include the following:

- System and subsystem requirements (I)
- System design and COTS (I)
- Unit test plans (I)

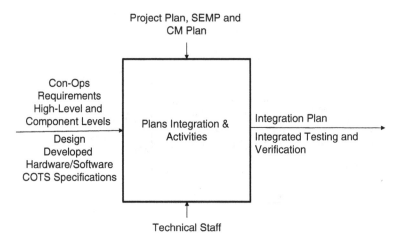

Fig. 2.11 IDEF0 activity diagram representation (partial) for system integration

- Software/hardware development plans (O)
- Hardware and software components (O)
- ILS documentation (O)

2.3.7 Systems Integration and Verification

In this stage, the software and hardware components are individually verified and then integrated to produce higher-level assemblies or subsystems. The scopes of integration and verification are illustrated in Figs. 2.11 and 2.12. The objectives are listed below:

- Integrate and verify the system and confirm that all requirements and constraints have been satisfied
- To confirm that interfaces have been implemented

  The sources of inputs (I) and outputs (O) include the following:

- System requirements document (I)
- High-level and detailed design specifications (I)
- Hardware and software units (I)
- Integration plan (I)
- Verification and acceptance plans (I)
- Master integration plan (O)
- Master verification plan (O)
- Integration analysis and verification results (O)
- Required corrective actions (O)

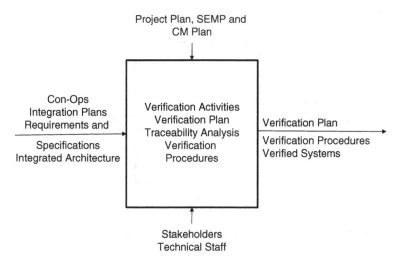

Fig. 2.12 IDEF0 activity diagram representation (partial) for system verification

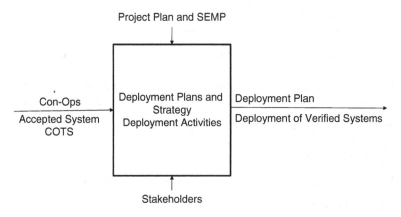

Fig. 2.13 IDEF0 activity diagram representation (partial) for initial deployment

2.3.8 Initial Deployment

At the end of this stage, the system is installed in the operational environment and transferred from the project development team. The objective is successfully transferred from the system to the stakeholder. The scope is illustrated in Fig. 2.13.

The sources of inputs (I) and outputs (O) include the following:

- Plan for system installation and transition (I)
- Prepare, deliver, and install the system (I)

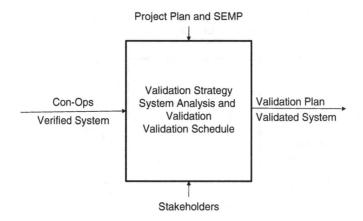

Fig. 2.14 IDEF0 activity diagram representation (partial) for initial deployment

- Perform acceptance tests to verify the system (I)
- Training materials (O)
- Delivery and installation plan (O)
- Transition plan and testing (O)
- Operations and maintenance plan and procedures (O)

This ninth stage of the systems engineering "V" process is systems validation shown in Fig. 2.14.

The objective of this stage is to validate the system and ensure that it meets the operational requirements of the stakeholders. Key activities are listed below:

- Update the validation plan and develop procedures
- Validate the system
- Document validation results

The sources of inputs (I) and outputs (O) include the following:

- Concept of operations (I)
- Installed and operational verified system (I)
- Master validation plan and procedures (O)
- Validation results
- Validated system (O)

2.3.9   Operations and Maintenance

The next stage of the SE cycle is the operations and maintenance. After the completion of the systems validation and the satisfaction of the ITS system by the customer, the system operates in a steady-state manner with maintenance

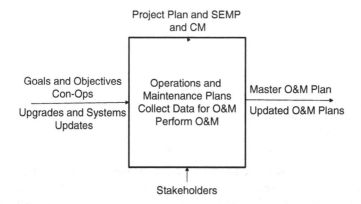

Fig. 2.15 IDEF0 activity diagram representation (partial) for operations and maintenance

routinely performed. The objective is to maintain the system over its operational life. The key activities are listed below:

- Establish and maintain operations and maintenance procedures
- Provide user and maintenance activity support
- Collect system operational data and maintain configuration control of the system
- Change or upgrade the system

   Figure 2.15 illustrates the scope of this stage.
   The sources of inputs (I) and outputs (O) include the following:

- Systems operational requirements (I)
- Systems maintenance requirements (I)
- Operations and maintenance plan and procedures (I)
- Training materials (I)
- Systems updates and changes (I)
- System performance reports (O)
- Systems maintenance records (O)
- Updated operations and maintenance procedures (O)
- Identified defects and recommended enhancements (O)

2.3.10 Retirement and Replacement

This is the final stage of the "V" process. In this stage, the system is frequently inspected to ensure efficient operational conditions. If the system's useful life has been attained and the system is no longer functional, it is then retired and replaced. The key activities are: (1) plan system retirement and (2) deactivate, remove, and dispose (recycle) system. The objective is to remove and retire the system.

The scope is illustrated in Fig. 2.16. The sources of inputs (I) and outputs (O) include the following:

- System requirements (I)
- Retirement/disposal requirements (I)
- Service life of the system and components (I)
- Service life of the components (I)
- System performance measures (I)
- Maintenance records (I)
- System retirement plan and procedures (O)

Conclusively the benefits of the systems engineering methodology in ITS systems have been summarized using the noted "V" process. For ITS to be designed and developed efficiently, it is advisable to adopt systems engineering methods and theories. Table 2.1 summarizes the application of systems engineering "V" process for design and development of ITS.

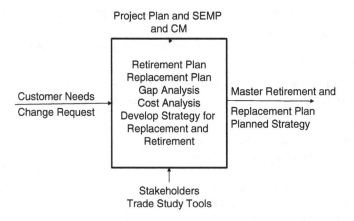

Fig. 2.16 IDEF0 activity diagram representation (partial) for replacement and retirement

Table 2.1 Detailed descriptions of "V" model stages [11]

Using the regional ITS architecture	The portion of the regional ITS architecture that is related to the project is identified. Other artifacts of the planning and programming processes that are relevant to the project are collected and used as a starting point for project development. This is the first step in defining ITS project
Feasibility study/concept exploration	A business case is made for the project. Technical, economic, and political feasibility is assessed; benefits and costs are estimated; and key risks are identified. Alternative concepts for meeting the project's purpose and need are explored, and the superior concept is selected and justified using trade study techniques

(continued)

Table 2.1 (continued)

Concept of operations	The project stakeholders reach a shared understanding of the system to be developed and how it will be operated and maintained. The concept of operations (ConOps) is documented to provide a foundation for more detailed analyses that will follow. It will be the basis for the system requirements that are developed in the next step
System requirements	The stakeholder needs identified in the concept of operations are reviewed, analyzed, and transformed into verifiable requirements that define what the system will do but not how the system will do it. Working closely with stakeholders, the requirements are elicited, analyzed, validated, documented, and baselined
System design (high-level and detailed design)	A system design is created based on the system requirements including a high-level design that defines the overall framework for the system. Subsystems of the system are identified and decomposed further into components. Requirements are allocated to the system components, and interfaces are specified in detail. Detailed specifications are created for the hardware and software components to be developed, and final product selections are made for off-the-shelf components
Software/hardware development and testing	Hardware and software solutions are created for the components identified in the system design. Part of the solution may require custom hardware and/or software development, and part may be implemented with off-the-shelf items, modified as needed to meet the design specifications. The components are tested and delivered ready for integration and installation
Integration and verification	The software and hardware components are individually verified and then integrated to produce higher-level assemblies or subsystems. These assemblies are also individually verified before being integrated with others to produce yet larger assemblies, until the complete system has been integrated and verified
Initial deployment	The system is installed in the operational environment and transferred from the project development team to the organization that will own and operate it. The transfer also includes support equipment, documentation, operator training, and other enabling products that support ongoing system operation and maintenance. Acceptance tests are conducted to confirm that the system performs as intended in the operational environment. A transition period and warranty ease the transition to full system operation
System validation	After the ITS system has passed system verification and is installed in the operational environment, the system owner/operator, whether the state DOT, a regional agency, or another entity, runs its own set of tests to make sure that the deployed system meets the original needs identified in the concept of operations

(continued)

Table 2.1 (continued)

Operations and maintenance	Once the customer has accepted the ITS system, the system operates in its typical steady state. System maintenance is routinely performed and performance measures are monitored
	As issues, suggested improvements, and technology upgrades are identified, they are documented, considered for addition to the system baseline, and incorporated as funds become available. An abbreviated version of the systems engineering process is used to evaluate and implement each change. This occurs for each change or upgrade until the ITS system reaches the end of its operational life
Retirement/replacement	Operation of the ITS system is periodically assessed to determine its efficiency. If the cost to operate and maintain the system exceeds the cost to develop a new ITS system, the existing system becomes a candidate for replacement. A system retirement plan will be generated to retire the existing system gracefully

Chapter 3
Methodologies and Tools for Evacuation Planning

Abstract Evacuation planning involves an iterative process to identify the best routes and to estimate the time required to evacuate the areas at risk. Methods typically used for evacuation planning includes *analytical models* to express the route choice and traffic propagation using mathematical equations and *simulation-based models* to describe the traffic conditions and pattern based on a set of rules. This chapter outlines the evacuation planning process and introduces the analytical representation of the evacuation models. A state-of-the-art review for some analytical and simulation models and methodologies for evacuation planning is also presented in this chapter.

3.1 Introduction

Urban evacuation can be viewed as a process in which evacuees are moved from dangerous areas to safe zones utilizing transportation resources. This massive movement of residents typically exceeds normal demands on transportation resources and thus requires careful planning and optimization. The greatest risk of a poor evacuation plan is that people may lose their lives if they are not given the chance to evacuate on time. Another less severe consequence is a massive number of people trapped for hours on the road. Figure 3.1 illustrates the elements involved in evacuation planning.

3.2 Evacuation Modeling Methodology

Recently, evacuation planning and modeling have increasingly attracted interests among researchers as well as government officials. This is probably due to the fact that south eastern states are under increasing threat of hurricane attacks every year.

A. Naser and A.K. Kamrani, *Intelligent Transportation and Evacuation Planning:*   37
A Modeling-Based Approach, DOI 10.1007/978-1-4614-2143-6_3,
© Springer Science+Business Media, LLC 2012

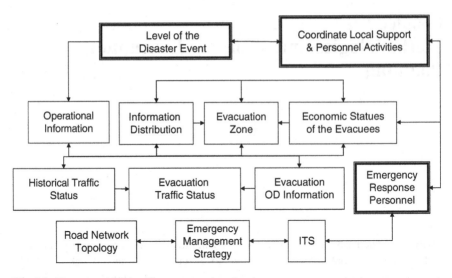

Fig. 3.1 Elements within intelligent evacuation planning

The catastrophic loss of human lives after hurricane Katrina and the massive evacuation "nightmare" before hurricane Rita in 2005 have also initiated several research efforts for the development of evacuation planning strategies and techniques. These research efforts use the latest achievements in a number of various disciplines, such as operations research, simulation, traffic and civil engineering, process control, psychology, architecture, physics, and many others. In general, methods used for evacuation planning include the following:

- *Analytical models* to express the route choice and traffic propagation using mathematical equations.
- *Simulation-based models* to describe the traffic conditions and pattern based on a set of rules.

Evacuation planning involves an iterative process to identify the best routes and to estimate the time required to evacuate the areas at risk. Common steps in this methodology are the following:

- Identify the region to be evacuated. Regions are defined as subsets of the entire area at risk and based on the location and direction of moving threats.
- Identify the demand (in vehicles) over each region to be evacuated. Demand that the population be subdivided into permanent residents, employees who work in the area at risk, and transients who are passing through the area or staying in the area temporarily. This demand is distributed to zonal centroids, which describe the changes in population density over the area.
- Identify safe region boundaries or shelters locations.
- Identify transportation network (supply) topology.
- Estimate roads link capacities based on field survey observations and on scenario-based weather conditions.

- Define the candidate destinations on the periphery of the region for each origin centroid. These destinations represent the points where network links cross the outer boundaries of the region.
- Compute the optimal routing of evacuation trips out of the region via the specified destination nodes.
- A traffic simulation model is then applied to simulate the movement of vehicles during the course of the evacuation. The model should explicitly describe traffic conditions in the saturated flow regime to account for congestion effects.
- Review simulation results to determine the need for traffic management to support the evacuation movements.

The ultimate goal of any evacuation planning effort is to minimize the number of casualties against any possible threats. One important factor leading to minimizing the number of causalities is total evacuation time, i.e., the time from the start of evacuation until the time the last evacuee reaches safety (time to clear the network). Obviously, minimizing total evacuation time can help in eliminating the risk of exposure to threats as well as minimizing the hardship endured in the evacuation process. Other objectives can be sought as well during evacuation planning including minimizing average evacuation time per evacuee, minimizing evacuation distance traveled, and maximizing number of evacuees reaching safety.

Evacuating a large population due to natural or man-made emergency situations is a complicated and costly task. Over the years, many researchers have been developing models for emergency planning and evacuation. These include *traffic simulation* and *route-schedule planning*, but none in an integrated form. The focus is on developing a model that finds optimal routes and schedules for evacuees in order to minimize total evacuation time. This can be called the evacuation problem which is generally stated as: *Given a set of danger regions, safety regions (represented by their centriods), and number of evacuees located in danger areas, what are the optimal routes and schedules for these evacuees to reach safety in such a way that total evacuation time is minimized.* Evacuation routing belongs to a class of network optimization problems called dynamic routing. A routing problem, in general, concerns with finding the "best" path(s) from a source to a destination that satisfy a specified objective. Other areas that involve dynamic routing include logistics and distribution, communications, and urban traffic planning. All these areas, in addition to evacuation planning, face a common problem of finding the best path for a commodity of flow in a network between the source and destination. Defining the best path depends on the application. In the evacuation area, it is the path(s) that result in the minimum evacuation time. Most previous analytical works on dynamic routing in general, and evacuation planning especially, have a major shortcoming: it assumes that travel time is constant. Even though this assumption simplifies calculations and facilitates using efficient solution methodologies, it can produce impractical and ineffective results. Almost everyone have experienced congestion on highways and noticed the difference in travel time between a congested and an empty road. Therefore, it is more realistic to assume that travel time depends on the flow. In this book, a new dynamic routing model is presented for evacuation planning that assumes load-dependent travel time. Recent works on the evacuation problem are mainly following two approaches: analytical models and simulation [13, 14]. Analytical models are primarily based on dynamic network optimization, and simulation models are mainly utilizing cellular automata principles. Both approaches have advantages and shortcomings, and the purpose of the methodology is to improve some of these shortcomings.

The main advantage of most analytical models is the capability of finding the "best" routes to evacuate, whereas the main drawback is that they assume constant travel time (speed). Under constant speed assumption, the evacuation problem can be solved using existing network optimization literature. A network is a graph $G = (N, A)$ with a set of nodes, N, and a set of arcs, A. An arc (i, j) connects two nodes i and j in the graph such that $i, j \in N$. In transportation context, roads are represented by arcs, and intersection, ramps, and exists are represented by nodes. Each arc (i, j) can have two parameters: travel time (τ_{ij}) and capacity (u_{ij}). Each node i can have also a parameter, $b(i)$, which represents the supply or the demand that has to be satisfied at that node. This supply equals the difference between total flow entering the node and the total flow leaving the node. The following is an example of a simple network.

Several network optimization problems appear in the literature that can fit evacuation planning according to their objective:

- A typical objective in evacuation is to minimize total evacuation time (time to clear the network). This objective is known in the network optimization literature as the quickest flow problem, and it arises when the number of evacuees that need to be cleared can be estimated.
- When the number of initial network occupants is unknown, the objective is to move as many evacuees as possible within specified time horizon. This objective is equivalent to a maximum dynamic flow problem, where the solution is achieved by sending as much flow as possible from the danger areas to the safe zones within time horizon.
- A more conservative objective is to maximize the number of evacuees reaching safety not only within the overall time horizon but within every smaller time period in the horizon. The same problem is known as the universal maximum flow or the earliest arrival problem.

To further illustrate the problem of constant travel time assumption, consider the network shown in Fig. 3.2. Assume that we are trying to decide how we can send the supply of 15 from node "S" to node "t" in the minimum amount of time. It is not hard to find that sending a flow of 10 along path S–1–t starting at time 0 and a flow of 5 along the same path starting at time 1 results in the minimum evacuation time of 4 (the second parameter on the arc represents the maximum amount of flow that can enter the arc in one time period). For the same network, it can be shown below, when considering travel time as a function of flow (density), another solution may be superior to the optimum solution above. Several speed–density models have been developed in traffic theory literature [15]. The goal here is neither to pick "the"

Fig. 3.2 An example of a
network

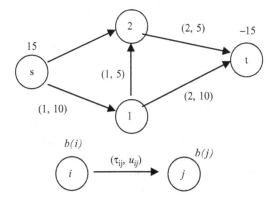

Table 3.1 Input parameters for Example

Arc	Distance, D (miles)	Free flow speed, v_f (miles/h)	Jam density, k_j (veh./miles)	Capacity U (veh./h)
$(S, 1)$	1	1	10	10
$(1, 2)$	1	1	5	5
$(1, t)$	2	1	10	10
$(2, t)$	2	1	5	5

perfect model nor to assess these models, rather to pick a model for demonstration. A famous model is the linear Greenshield's equation relating speed (v) and density as follows [15]:

$$v = v_f(1 - k/k_j),\tag{3.1}$$

where k is the current density, k_j is the jam density, and v_f is the free flow speed. Let us now assume that the following parameters are for the arcs in Fig. 3.2. Assuming that a flow entering an arc will travel at the free flow speed if the arc is empty and it will travel at a speed defined by (3.1) if it is occupied. Going back to the previous solution, a flow of 10 starting at time "0" will find all arcs empty and thus will travel at the free flow speed (i.e., it will arrive as scheduled). However, the flow of 5 starting at time 1 will arrive node 1 at time 2 and will find arc (1, t) still occupied by the previous flow (Table 3.1).

Using (3.1), the speed at which this flow will travel is calculated as follows:

- Current density, $k = 10/2 = 5$ veh./miles
- Current speed, $v = 1(1 - 5/10) = 0.5$ miles/h with arc (1, t) travel time = 2/0.5 = 4 h

Considering the new travel time means that total evacuation time is 6 h. A better alternative schedule is by sending a flow of 10 along path S–1–t starting at time 0 and a flow of 5 along path S–1–2–t starting a time 1. Total evacuation time now is 5 h.

From the above discussion, an optimum solution under constant travel assumption may not remain the best alternative if we allow, more realistically, travel time to vary according to the flow on the arc. Furthermore, there is no guarantee that constant travel time solution will even perform "good" and close to the optimum.

In the worst-case scenario, this solution may result in severe congestion and perform worse than most alternative solutions. Therefore, this assumption can result in inaccurate, impractical, or ineffective results. Consequently, building a routing model that eliminates constant travel time assumption can significantly improve the quality and reliability of solutions. In order to achieve such a model, we have to build a model that can perform two tasks: how to find the best routes for evacuation and how to predict travel time as a function of flow.

Simulation models can predict travel time as a function of flow but they act merely as a tool for evaluating different scenarios and devising recommendations. The main drawback in simulation models is that they do not have the capability of identifying the "best" routes (optimization). This observation, actually, sparked the following idea. If we are able to build a model that integrates an optimization and a simulation routine into one algorithm, then significant improvements to evacuation routing can be achieved. In this book, an integrated model for dynamic routing and road traffic simulation is presented. The goal is to minimize total evacuation time while considering travel theories, mainly load-dependent travel times. The main idea is to use a simulation routine to propagate the flow along the arcs. Once they reach the head node, an optimization routine selects the best routes to direct the flow out of the head node. The general framework is as follows [16]:

- *Developing a traffic simulation model.* A traffic simulation model is developed that can propagate a flow along the arcs and predict travel time as a function of arcs load. The research relies on existing civil and traffic engineering literature to build this model.
- *Developing an optimization model.* An optimization model is developed to minimize total evacuation time. It is based on network optimization formulations and principles.
- *Integrating both models into one algorithm.* The main idea is to discretize the time space and to solve both routines at each time step. The challenge remains as how to link an instance of the problem at time T to the instance at time $T + 1$, and how to maintain an equivalent problem of the original network at each time step. These challenges are solved and explored further in this book.
- *Evaluating the model.* The main theory in this book is that a routing solution that is obtained by assuming constant speed may produce, if applied in real-life settings, impractical and ineffective results. The new router should outperform and overcome this shortcoming. To test this theory, several case studies are developed. Each case study is solved first by using a constant speed (travel time) algorithm. This solution is then simulated through the case study using the developed traffic simulation model and the resulted evacuation time is recorded. For the same case study, the new router is used to find the best evacuation routes and the resulting evacuation time is recorded and compared against constant travel time result. The system's architecture is illustrated in Fig. 3.3.

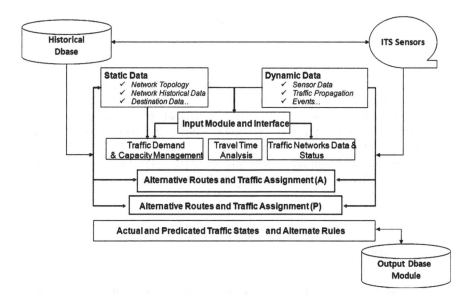

Fig. 3.3 Intelligent evacuation system's architecture

The evacuation research has been studied by scholars in different disciplines. These science disciplines include civil engineering, industrial engineering, physics, and even psychology and social science. A main goal in evacuation planning is to minimize total evacuation time. Several models, however, exist in the operation research literature for this objective depending on the desired level of abstraction, and consequently, the level of complexity. Different classifications can be established for these models as well based on solution methodology, scope, and input parameters. Three distinct classifications can be identified in evacuation modeling research: macroscopic analytical models, microscopic simulation models, and traffic assignment models.

Based on the scope, evacuation models can be classified into macroscopic and microscopic models. Some evacuation problems consider evacuees movement as a homogenous flow. These models are called macroscopic models where interactions among individual evacuees are ignored. As a benefit of this simplification, most macroscopic models are solved using mathematical or analytical techniques. Analytical models are based mainly on network optimization formulations and few integer programming models. These models suffer, however, from some loss of accuracy as individual behavior may influence total evacuation time. Other evacuation models, called microscopic models, consider individual behavior, and interactions among evacuees. The resulted level of complexity, however, discourages using analytical techniques to solve the problem. These problems are modeled and solved mainly using simulation approach. Simulation models are increasingly based on cellular automata.

Traffic assignment models are found in the civil engineering literature. They aim at evaluating transportation network performance against projected traffic demand between specified origin and destination pairs. The main difference between these models and most of analytical evacuation models is that they allow travel time to be a function of road flow and density. Flow can be defined as vehicle per unit of time and density as vehicle per unit distance. Most analytical models that are based on network optimization models simply ignore this characteristic of traffic flow and assume constant travel time. Even though this simplification allows using efficient network optimization algorithms, it suffers from lack of accuracy and low solution reliability. This section presents the background of evacuation problems and summaries related research found in the literature. The materials presented about analytical models are not intended to be self-contained and the reader is expected to be familiar with basic principles about network optimization [1].

3.3   Macroscopic Models

Macroscopic evacuation models ignore individual behavior for evacuees and consider evacuation movement as a homogenous flow. The theory behind solving these problems comes mainly from network optimization research.

Classical network optimization problems fit the structure and objective of the evacuation problem. A common element in network optimization problems is the network representation. A network is a graph $G = (N, A)$ with a set of nodes, N, and a set of arcs, A, such that if an $arc(i,j) \in A$ then $i \in N$ and $j \in N$. Each node and each arc can have several input parameter. Each node, for example, can have a parameter of a supply or a demand that has to be satisfied at that node. Arcs can have multiple parameters that include arc capacity, arc travel time, and cost per unit flow on that arc. In urban evacuation context, the transportation structure is represented by a network where arcs represent roads and nodes represent intersections.

Previous work on mathematical evacuation (macroscopic) modeling can be generally divided based on using static or dynamic network formulations. Classical network optimization problems include shortest path problem, maximum flow problem, and minimum cost flow problem. These problems are considered static because they do not include a time dimension, i.e., a solution to the problem is not a function of time. When considering the time dimension, network problems are categorized as dynamic. Related dynamic problems include maximum dynamic flow problem, quickest flow problem, and earliest arrival (universally max flow) problem.

Dynamic network optimization is more appropriate for evacuation planning for the added accuracy when considering the time dimension. Therefore, the focus in this section is more toward dynamic network theories and algorithms as follows: first, static and dynamic problems are introduced and formulated. Detailed discussion is then presented about dynamic network theories. Finally, a summary of literature review about previous evacuation models is discussed.

3.3.1 Static Network Optimization

The shortest path problem. A classical and important problem in network optimization is the shortest path problem. In this problem, a graph $G = (N, A)$ with an arc length of c_{ij} associated with each arc (i, j) in A and two designated nodes called start node "s" and destination nodes "t", the goal is to find the shortest path from node s to node t. The length of the path is defined as the sum of the length of all arcs in the path. There is extensive literature addressing the shortest path problem [17]. The famous Dijkstra's algorithm is an efficient method to solve many instances of the problem.

Maximum flow problem. The goal in a maximum flow problem is to send as much flow as possible from a source node to a sink node while satisfying arcs capacities and flow conservation constraints. If we define X_{ij} as the flow from node i to node j along arc (i, j), and U_{ij} as the capacity of arc (i, j), the maximum flow problem can be formulated as follows:

$$\text{Max } v,$$

s.t.

$$
\begin{aligned}
&\sum_{j:(i,j)\in A} X_{ij} - \sum_{j:(j,i)\in A} X_{ji} = 0, \quad \forall i \in N - \{s,t\}, \\
&\sum_{j:(i,j)\in A} X_{ij} - \sum_{j:(j,i)\in A} X_{ji} = v, \quad \text{for } i = s, \\
&\sum_{j:(i,j)\in A} X_{ij} - \sum_{j:(j,i)\in A} X_{ji} = v, \quad \text{for } i = t, \\
&0 \leq X_{ij} \leq U_{ij}, \quad \forall (i,j) \in A.
\end{aligned}
\tag{3.2}
$$

The minimum cost problem. Given a network $G = (N, A)$:

- For each node $i \in N$, a number $b(i)$ is associated that indicates if supply $(b(i) > 0)$ or demand $(b(i) < 0)$ is required at node i.
- A cost C_{ij} and capacity U_{ij} is associated for each arc $(i,j) \in A$.

The goal is to find a feasible flow that satisfies supply and demand with the minimum cost. The formulation is as follows:

$$\text{Min} \sum_{(i,j)\in A} C_{ij} X_{ij},$$

s.t.

$$
\begin{aligned}
&\sum_{j:(i,j)\in A} X_{ij} - \sum_{j:(j,i)\in A} X_{ji} = b(i), \quad \forall i \in N, \\
&0 \leq X_{ij} \leq U_{ij}, \quad \forall (i,j) \in A.
\end{aligned}
\tag{3.3}
$$

3.3.2 Dynamic Network Optimization

The max dynamic flow problem. In a static max flow problem, the only parameter is arcs capacities and the goal is to send as much flow as possible from a source node to a sink node. Dynamic flow problems, on the other hand, have two parameters associated with each arc: travel time and capacity. In a max dynamic flow problem, the goal is to send as much flow as possible from the source node to the sink *within* a specified time horizon T. Let

- τ_{ij} be the time to traverse arc (i, j) from node i to node j, and u_{ij} be the capacity of arc (i, j)
- $X_{ij}(t)$ be the flow along arc(i, j) that leaves node i at time t and arrives node j at time $(t + \tau_{ij})$
- U_{ij} be the capacity of arc (i, j)
- U_{ii} be the maximum flow that can be held over at node i
- $S =$ the set of all source nodes in the static network G
- $D =$ the set of all sink nodes in the static network G
- $D =$ a super sink in the dynamic network G^T

The max dynamic flow can be then formulated as follows:

$$\text{Max} \sum_{t=0}^{T} \sum_{i \in D} X_{id}(t),$$

s.t.

$$X_{ii}(t+1) - X_{ii(t)} = \sum_{\substack{j \in N \\ (j,i) \in A}} X_{ji}(t - \tau_{ji}) - \sum_{\substack{j \in N \\ (i,j) \in A}} X_{ij}(t), \ \forall i \in N, \ t \in \{0, 1, ..., T\},$$

$$(3.4)$$

$$X_{ii}(0) = 0, \ \forall i \in N,$$
$$X_{ii}(t) = 0, \ \forall i \in S \cup D, \ t \in \{1, ..., T\},$$
$$0 \leq X_{ij}(t) \leq U_{ij}, \ \forall (i,j) \in A, \ t \in \{0, 1, ..., T - \tau_{ij}\},$$
$$0 \leq X_{ii}(t) \leq U_i, \ \forall i \in N, \ t \in \{0, 1, ..., T - 1\}.$$

3.3.3 Solving Max Dynamic Flow Problems

Time expanded network. A dynamic network can be considered as being composed of multiple copies of snapshots of a static network at different points of time; one at each point of time [20]. Using this view, the static $G = (N, A)$ can be transformed into a new graph $G^T(N^T, A^T)$ called time expanded network as follows:

$$N^T : \{i(t) : i \in N; \quad t = 0, 1, 2, ..., T\}, \tag{3.5a}$$

$$A^T = \{A_M^T \cup A_H^T\}, \tag{3.5b}$$

$$A_M^T = \{i(t), j(t + \tau_{ij}) : (i,j) \in A; \quad t + \tau_{ij} \leq T; \quad t = 0, 1, 2, ..., T\}, \tag{3.5c}$$

$$A_H^T = \{i(t), i(t + 1) : i \in N; \quad t = 0, 1, 2, ..., T - 1\}. \tag{3.5d}$$

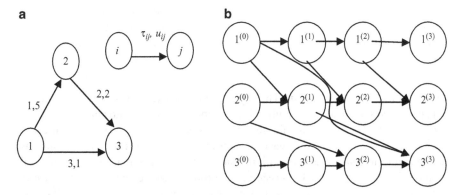

Fig. 3.4 (a) Static network and (b) time expanded network with $T = 3$

The set of arcs A^T is composed of movement arcs and hold over arcs. For example, the static network shown in Fig. 3.4 is transformed to the time expanded network.

Using this approach, dynamic flow problems can be solved by applying known static network algorithms to the time expanded network. A drawback for this method is that problem size is exponential with the number of time periods, T, and thus makes it inefficient methodology for real-life problems that are usually large. If the static model has n nodes and m arc, and the dynamic model has T time horizon, then the upper bound on the number of arcs in the dynamic model is $(n + m)T$ and the upper bound on the number of nodes is $n(T + 1)$.

Temporary repeated chain approach. Ford and Fulkerson [21] showed that some instances of maximum dynamic flow problem can be solved through solving single static min cost problem. This approach, called the temporary repeated chain approach, is achieved by finding first a static flow that minimizes a cost function. This flow is then decomposed into chain flows from source to sink. A max dynamic flow is generated by repeating each decomposed flow from source to sink starting at time 0 and continuing at each time step as long there is enough time for the flow to reach the sink before time T.

A chain flow $f = (v, P)$ is a static flow of value v from source node to the sink along path P. The length of chain f is $\tau(f)$, where $\tau(f) = \sum_{(i,j) \in P} \tau_{ij}$. We call $F = \{f_1, f_2, \ldots, f_k\}$ chain decomposition of static flow X if $\sum_{f \in F} v(f) = X$. Each chain flow induces a finite horizon dynamic flow (bounded by time T) by sending v units of flow from source to sink every time step from time 0 till time $T - \tau(f)$. The collection of all flows induced this way is a feasible dynamic flow, F^T, that is called the temporary repeated chain flows. The value of this dynamic flow is independent of the chain decomposition and depends only on the static flow [24].

$$F^T = \sum_{(i,j) \in P} (T - \tau(f) + 1)|f|,$$
$$= (T + 1)|X| - \sum_{(i,j) \in A} \tau_{ij} X_{ij}. \tag{3.6}$$

The first term above is constant; therefore, the dynamic flow F^T can be maximized by minimizing the second term. This can be done by introducing a new arc from the sink to the source with cost equal to, $-(T+1)$, and infinite capacity, and solving the resulted min cost circulation problem where arc traversal time is the arc cost.

Ford and Fulkerson algorithm. The algorithm constitutes two routines: Routine I constructs the static flow and Routine II decomposes this flow into chain flows from source to sink. The max dynamic flow can then be induced by starting each chain flow at time 0, and repeating each at subsequent time steps as long as there is enough time to reach the sink within time T. Routine I is mainly a primal dual method for the capacitated transshipment problem where the static flow for T periods proceed from the one previously obtained for $T-1$ periods. The Routine involves an iterative process which has an output integral static flow X_{ij} along with node potentials π_i for each $i \in N$ such that

$$\pi_s = 0, \quad \pi_t = T+1, \quad \pi_i \geq 0, \quad \forall i \in N,$$
$$\text{if } \pi_i + \tau_{ij} > \pi_j \text{ then } X_{ij} = 0, \qquad \forall (i,j) \in A,$$
$$\text{if } \pi_i + \tau_{ij} < \pi_j \text{ then } X_{ij} = U_{ij}, \qquad \forall (i,j) \in A.$$

An arc (i, j) is called admissible if $\pi_i + \tau_{ij} = \pi_j$. During the algorithm, the possible states for any node are unlabeled, labeled but unscanned, or labeled and scanned.

Routine I:

Initialization
$\pi_i = 0$ for all $i \in N$, $X_{ij} = 0$ for all $(i, j) \in A$;
For all $i \in N$ do   Pred [i] = -1 , $\delta[\,i\,] = \infty$;
Label= {}, scanned = {};
Pred [s] = t$^+$; Label = {s}
While $\pi_t < T+1$
While   more labeling is possible   AND   node t is unlabeled
         For all labeled unscanned node i do
                 For all $j \in A(i)$, do
                         IF node j is unlabeled AND arc(i,j) is admissible
                         AND $X_{ij} < U_{ij}$ THEN
                             Pred[$\,j\,$] = i^+, $\delta[\,j\,]$ =   min{ $\delta[\,i\,]$, U_{ij} - X_{ij} };
                             Label = Label + {j}
                         IF node j is unlabeled AND arc(j,i) is admissible
                         AND $X_{ji} > 0$ THEN
                             Pred[$\,j\,$] = i^-, $\delta[\,j\,]$ =   min{ $\delta[\,i\,]$, X_{ji} };
                             Label = Label + {j}
                 End for
                 Scanned = Scanned + {i }
         End for

End While
If node t is labeled THEN
      Update flow
      Start from new generated flow and discard all labels
If no label is possible THEN
      For all $i \in N$ do
            If node is not labeled THEN π_i ++;
      End for
End if
End While

Routine II:

Pred $[\, s\,] = t$; $\delta[\, s\,] = \infty$; s is unscanned
While   more labeling is possible
      While   node t is unlabeled
            For all labeled unscanned node i do
                  For all $j \in A(i)$, do
                        IF node j is unlabeled AND $X_{ij} > 0$ THEN
                              Pred$[\, j\,] = i$, $\delta[\, j\,] = $ min$\{\ \delta[\ i\ ]$,$X_{ij}\ \}$;
                              Label = Label + $\{j\}$
                  End for j
                  Scanned = Scanned + $\{i\ \}$
            End for
      End While
      If node t is labeled THEN
            Update flow
      End if
End While

Earliest arrival flow (universal max flow). The earliest arrival flow simultaneously maximizes the amount of flow reaching the sink at every time step t and not only at the end of the time horizon T. An earliest arrival flow is obviously a max dynamic flow for the whole time horizon, but the vice versa is not necessarily true. The problem is formulated as follows:

$$\text{Max} \sum_{t=0}^{\Gamma} \sum_{i \in D} X_{id}(t), \quad \forall\, \Gamma \in \{1, 2, ..., T\},$$

s.t.

$$X_{ii}(t+1) - X_{ii(t)} = \sum_{\substack{j \in N \\ (j,i) \in A}} X_{ji}(t - \tau_{ji}) - \sum_{\substack{j \in N \\ (i,j) \in A}} X_{ij}(t), \quad \forall\, i \in N, \quad t \in \{0, 1, ..., T\}, \tag{3.7}$$

$$X_{ii}(0) = 0, \quad \forall\, i \in N,$$
$$X_{ii}(t) = 0, \quad \forall\, i \in S \cup D, \quad t \in \{1, ..., T\},$$
$$0 \le X_{ij}(t) \le U_{ij}, \quad \forall\, (i,j) \in A, \quad t \in \{0, 1, ..., T - \tau_{ij}\},$$
$$0 \le X_{ii}(t) \le U_i, \quad \forall\, i \in N, \quad t \in \{0, 1, ..., T - 1\}.$$

Wilkinson and Minieka showed how to modify the Ford and Fulkerson repeated chain approach to obtain an earliest arrival flow [22, 23]. The main idea in their algorithm is to solve the static min cost flow problem via the shortest augmenting path algorithm. The augmentations of this algorithm defines a chain decomposition $F*^T$ of $X*$ which is repeated to obtain the earliest arrival flow. Hoppe and Tardos presented a review for some polynomial time algorithms for evacuation problems [24]. They developed also an approximation polynomial algorithm for the earliest arrival flow. The algorithm is capacity scaling that scales upward. Usually, capacity scaling algorithms scale downward but in dynamic flow; a small capacity arc that is short might carry more flow than a large capacity arc that is long. The algorithm finds a minimum cost flow in a rounded static network via shortest augmenting paths algorithm. The chain decomposition defined by the sequence of augmentations is repeated to induce a dynamic flow. The algorithm sequentially rounds down the residual capacities by an increasing scaling factor Δ. The rounding guarantees that the algorithm is polynomial in n, log U, and ε, where $\varepsilon > 0$ is an error parameter.

Quickest flow problem (the evacuation problem). The quickest flow problem is strongly related to the max dynamic flow problem [25, 51]. The goal here is to send a specified amount of flow from the source node to the sink node with the minimum amount of time. The problem is formulated as follows (assuming "v" initial units of flow exist at the source node): Min $T(v) = \min\{T|v(T) \geq v\}$ where $v(T)$ is a maximum flow over time period T from maximum dynamic flow problem.

Lexicographic maximum dynamic flow. In the lexicographic maximum problem, there exist multiple sources, and these sources are ordered based on priority such that a source ranked one has the highest priority. The goal in this problem is, given a time horizon T, to maximize the flow leaving the sources to the sink according to the lexicographic order of sources. This means a solution that maximizes the flow leaving the source with the highest priority is better than any other flow regardless of the amount of flow leaving less priority resources. If two solutions have equal flow leaving the highest priority source, they are compared against next highest priority source regardless of the rest of sources, and so on.

Hope a Tardos proposed a polynomial time algorithm that computes a chain decomposition F, such that the resulted dynamic is a lexicographically maximum dynamic flow. The algorithm is briefly as follows: a super source S node is added with infinite capacity arcs (t, S) connecting the sink with $- (T + 1)$ transit time, and (S, s_i) arc connecting each source s_i with zero transit time. Let this graph be denoted as G^k. A minimum cost circulation g^k in static network is calculated using transit times as costs. Then, in each iteration i, G^{i+1} is modified by deleting the arc (S, s_{i+1}) to create G^i. A minimum cost maximum flow f^i is calculated from S to s_{i+1} in the residual graph of the flow g^{i+1} in G^i. The resulted minimum cost flow is then added to the flow g^{i+1} to obtain the flow g^i. Each flow f^i yields standard chain decomposition that induces a maximum dynamic flow.

Applying the shortest path formulation in evacuation was done by Fahy [18]. He developed a model, Exit89, for evacuation of large building. The model briefly moves evacuees along a calculated shortest path from each building location to safety; or along a predetermined safe route defined by the user of the model. It can consider delays by blocked area due to smoke by recalculating exit routes until new blockage occurs or everyone reaches safety.

Yamada [19] used a network flow approach to model city emergency evacuation. In this approach, residential areas and places of refuge are modeled as nodes and roads in-between as arcs. The model uses two methods for evacuation. The first method assigns each residential area to the nearest places of refuges by calculating shortest path to an artificial node. This artificial node is connected to every place of refuge with zero distance arcs and the shortest path is calculated from this node to every residential area. This method minimizes individual as well as total distance to evacuate. The second method solves a min cost flow problem to assign residential areas to places of refuges while considering capacity limits on these places of refuges. Nodes that represent resident areas will have a demand that need to be satisfied and refuge places nodes will have capacities to be respected. The model adds two artificial nodes, a source node and a sink node. Arcs are added between the source node and all resident areas with a capacity equal to the demand and zero cost; and arcs between sink nodes and all refuge places with capacity equal to refuge capacity and zero cost. The original arcs will have infinite capacity and cost equal to distance between nodes in the graph.

Cova and Johnson proposed a mixed integer linear program for optimal evacuation routing [20]. The model is an extension to a min cost flow problem where constraint added with binary variables. The primary objective is to route vehicles to their closest evacuation zone exist. A secondary objective (constraint) is to minimize the number of intersection merging and crossing conflict.

Chalmet et al. proposed a dynamic flow model for building evacuation [26]. They constructed a time-expanded dynamic model for a static model of a building and denoting copy "0" of each source node as new source and copy "T" of each sink as new sink. The model is solved to minimize the average evacuation time by using the turnstile costing approach. The turnstile costing approach assigns a cost of "t" to each unit of flow (person) passing through an exit at the end of period t. In other words, if D is the set of sink nodes in the static network and $d*$ is the super sink node in the dynamic network, the turnstile cost is defined as follows,

$$c(i(t), j(t + \tau_{ij})) = \begin{cases} t, & i \in D, j(t + \tau_{ij}) = d \\ 0, & \text{otherwise} \end{cases} \quad \forall (i(t), j(t + \tau_{ij})) \in A^T, \quad (3.8)$$

The total turnstile cost represents the total number of time periods incurred by every flow existing in the building. Dividing the total cost by the number of evacuees gives the average time needed for one person to evacuate the building. This model can be solved as a static min cost flow model to minimize the average evacuation time.

Jarvis and Ratliff, using Theorem 1 below, showed that solving the min cost turnstile problem is equivalent to solving other dynamic objectives [43]. These objectives include the following:

- The number of evacuees existing in periods 1 through t is maximized for all values of t.
- The average number of periods an evacuee required to exist is minimized.
- The period in which the last evacuee exists is minimized.

Theorem 1 [27] Consider any feasible flow with a value v from s to t within time horizon T and the following three objectives:

(a) Maximize $v(P)$ for $P = 1, 2, \ldots, T$
(b) Minimize $\sum_{t=0}^{T} c_t(v(t) - v(t-1))$ where $v(-1) = 0$ and $c_1 < c_2 < \cdots < c_T$
(c) Minimize P such that $v(k + P) = v(P)$ for $k = 1, 2, \ldots, T - P$

Any feasible flow of v from s to t that satisfies either objective (a) or (b) also satisfies the other two. The reader is referred to [28] for a proof of this theorem. Hamacher and Tufekci used lexicographical optimization to consider multiobjective evacuation problem [28]. Minimizing total evacuation time while avoiding cyclic movements in a building and "priority evacuation" are treated as lexicographical minimum cost flow problem. They used the turnstile costing approach on time expanded network to represent and model buildings. The usual cost function is replaced with a vector of prioritized cost elements that can be compared lexicographically.

3.4 Microscopic Models

In a microscopic modeling, interactions between evacuees can be considered. The resulted complexity, however, prohibits using analytical models to solve the problem efficiently. Therefore, optimum solutions are not generally sought in microscopic modeling. Instead, a simulation approach can be used to evaluate current performance and to predict future performance under several proposed scenarios. Recently, simulation models for evacuation planning are increasingly based on cellar automata. Cellular automata are mathematical abstractions of some physical systems. These systems are characterized by discrete time and space, and in which physical quantities take on a finite set of discrete values [29]. In its simplest form, a cellular automaton consists of a one-dimensional array of cells or sites. The state of the cellular automaton is completely specified by the values of the variables at each cell. A cellular automaton evolves in discrete time steps with value of a variable at one cell being affected by its "neighbor" cell at the previous time step. Cellular automata were originally introduced in 1963 by Von Neuman and Ulm as a model of some biological self-production. Since then, they have been applied to model complex systems in physics, mathematics and computing, and biology. A major advantage of cellular automata is that, despite the simplicity of their

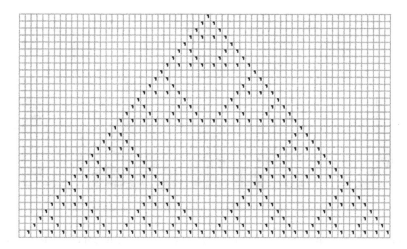

Fig. 3.5 Output of cellar automata according to the rule: 01011010

structure, cellular automata can produce behaviors of complex systems. For example, consider a cellular automaton that consists of one-dimensional array of cells. Each cell can have a value of 0 or 1 depending on the value of it immediate neighbor cells in the previous step according to the following rule that covers all eight possibilities of a cell and its two neighbors: 111/0 110/1 101/0 100/1 011/1 010/0 001/1 000/0, for example, if a cell has a value of 1 and its two neighbors are also 1, the next time step the cell value is 0. The evolution of this simple rule over time produces the following pattern (starting by 1 at time 0) (Fig. 3.5).

Nagel and Schreckenberg proposed a cellular automata model for freeway traffic [30]. The model is defined on a one-dimensional array of L sites, where each site can be either occupied by one vehicle or empty. Each vehicles has an integer velocity with values between 0 and v_{max}. At each time step, the following four consecutive steps are performed in parallel for all vehicles:

- *Acceleration.* If the velocity z of a vehicle is lower than v_{max}, and if the distance to the next car ahead is larger than $v + 1$, the speed is advanced by one, i.e., $v \rightarrow v + 1$.
- *Slowing down.* If a vehicle at site I sees the next vehicle at site $i + j$ (with $j \leq v$), it reduces its speed to $j - 1$, $v \rightarrow j - 1$.
- *Randomization.* With probability p, the velocity of each vehicle is decreased by one, $v \rightarrow v - 1$.
- *Car motion.* Each vehicle is advanced by v sites.

Klupfel et al. proposed a cellular automata model (Fig. 3.6) to simulate passenger ship evacuation [31]. The basic principle of their model is as follows:

- The ship floor plan is divided into equal quadratic cells with length of one side equals 0.4 m.
- The state of each cell at any time step is either empty or occupied by one person.

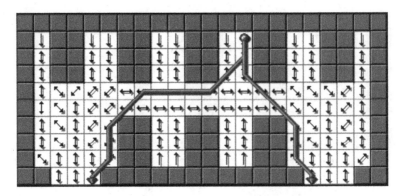

Fig. 3.6 A cellular automata model to simulate passenger ship evacuation [31]

- Individual persons can have parameters that describes different characteristics or abilities.
- The motion of people is defined by walking speed and direction that obeys universal laws.

The simulation is run by sequentially updating the position of individuals evacuating. The direction of movement is chosen such that it allows shortest escape and maximum speed. If the destination cell is occupied, then alternative cells are chosen by varying direction or speed. The update of individual position continues until the end of simulation run.

Farahmand [32] proposed a simulation model using WITNESS software for emergency evacuation. Model input variables include: population, local road capacities, number of registered cars, estimated tourist population, and warning accuracy based on last forecasts. The model provides theoretical evacuation times, actual time which considers delays, and predicted time if unidirectional traffic flow is followed away from the storm. Doheny and Fraser [33] proposed a software model for offshore personnel evacuation. This model takes into consideration human decision during evacuation process.

3.5 Traffic Theory and Traffic Assignment Models

Most of network formulations discussed before in this section assume that arcs travel time and capacities are constant. Considering real-life networks, urban traffic network, for example, requires further analysis to determine whether this assumption can produce useful and reliable results. When a road is empty except of one car, the driver of this car has the freedom to choose his own speed, and thus a constant travel time can be set. This speed is of course bounded by vehicle and road boundaries limit. As the degree of congestion increases, the desired speed of drivers may not be achievable. This is due to the fact that speed has to be reduced sometimes to maintain safe distance from other vehicles or due to the decrease in the ability to pass with condense traffic. Such interactions among vehicles, drivers, and road network structure minimize the accuracy and practicality of traffic models where travel time is assumed to be constant.

Traffic flow theories aim at mathematically modeling interactions among drivers, vehicles, and network infrastructure. Several models have been developed in this branch of science, such as the car following models and the continuum flow models [36]. Car following models examine the manner in which drivers follow other vehicles. These models form a bridge between microscopic and macroscopic traffic modeling where such interactions are ignored. Continuum flow models, on the other hand, consider traffic flow as a one-dimensional compressible fluid (macroscopic model). This analogy leads to two basic assumptions: traffic flow is conserved, and there is a one-to-one relationship that exists between speed and density or between flow and density. A simple continuum model consists of the conservation equation and the equation of state that specify the relationship between speed and density or between flow and density. Solving these equations, together with the basic traffic flow equation that states that flow is the product of speed by density, results in determining current traffic state and allows estimating travel time. An example of the equation of the state is the famous linear Greenshield's model relating speed (v) and density (k) as follows [19]:

$$v = v_f(1 - k/k_j),\tag{3.9}$$

where k_j is the jam density and v_f is the free flow speed. A more complex (higher-order) equation of state is the kinematics wave theory that is derived from fluid mathematics [36]. The theory assumes a functional relation between the traffic flow q and density n, also known as the fundamental diagram of traffic flow. Traffic propagates on a road according to this theory that is mainly based on the conservation of vehicles concept. The concept states that vehicles are neither created nor lost along a homogeneous link in a certain region of interest (Δx, Δt) if there exists a cumulative vehicle number function $N(x, t)$ in this region (Δx, Δt). The partial derivatives of $N(x, t)$ are the flow and density functions

$$q(x,t) = \frac{\partial N(x,t)}{\partial t}, \quad k(x,t) = \frac{-\partial N(x,t)}{\partial x}.\tag{3.10}$$

Provided that the function $N(x, t)$ exists in a certain region (Δx, Δt), as well as its first and second derivatives, the identity

$$\frac{\partial^2 N(x,t)}{\partial x \partial t} = \frac{\partial^2 N(x,t)}{\partial t \partial x}.\tag{3.11}$$

Together with the above two equations result,

$$\frac{\partial q(x,t)}{\partial x} + \frac{\partial k(x,t)}{\partial t} = 0. \tag{3.12}$$

It is out of the scope of this book to discuss the theory in further details; however, it may be beneficial to present a famous model that is considered as a discrete version of the kinematics wave theory. This model is called the cell transmission model (CTM) and it is discussed next.

3.6 Cell Transmission Model

The cell transmission model (CTM) was developed by Daganzo to model and simulate a road's traffic flow at a macroscopic level [36, 37]. This model is shown to be a discrete version of the kinematics wave theory. His model tries to improve the deficiencies in previous models that consider travel time as an increasing function of flow or as a function of incoming, existing, and exiting flows.

In the CTM, each arc is divided into cells where the length of each cell equals the distance traveled by a free-speed moving vehicle in one time click. Under light traffic conditions, all vehicles contained in cell i at time t in an arc will be assumed to advance to cell $i + 1$ at time $t + 1$, i.e., $n_{i+1}(t) = n_i(t)$. However, for crowded roads, traffic is slowed down by queuing from downstream bottleneck. To incorporate queuing, two parameters are introduced.

1. $N_i(t)$ is the maximum number of vehicles that can be present in cell i at time t. This will be the product of length of cell i by its jam density.
2. $Q_i(t)$ is the maximum number of vehicles that can flow into cell i when clock advances from t to $t + 1$. This will be the maximum flow or capacity for cell i.

$x_i(t)$ is the number of vehicles that can be advanced (the flow) from cell $i - 1$ to cell i in interval t to $t + 1$, and its equal the min of $n_{i-1}(t)$ number of vehicles present in cell $i - 1$ at time t, $Q_i(t)$ is the capacity flow into i for time interval t, and $N_i(t) - n_i(t)$ is the amount of empty space in cell i at time t multiplied by W/V (backward wave speed/free flow speed). Cell occupancy at time t equals the occupancy at time $t - 1$ plus inflow minus outflow, i.e.,

$$n_i(t+1) = n_i(t) + x_i(t) - -x_{i+1}(t), \tag{3.13}$$

where

$$x_i(t) = \min\{n_{i-1}(t), Q_i(t), \delta(N_i(t) - n_i(t))\} \text{ and } \delta = W/V. \tag{3.14}$$

Once $x_i(t)$ is determined using the above equation, $n_i(t)$ can be determined recursively from (3.13). Equations (3.13) and (3.14) above are used to predict and simulate flow along each arc.

3.7 Second-Order Macroscopic Traffic Models

A model like the CTM is considered a first-order model. The criterion here is that this model uses only one difference equation (3.13) to define the system state at each time step. First-order models, however, are reported in the literature to show poor transient behavior [38]. Second-order models use two difference equations to define the system. In addition to the number of cell occupants, the average speed in any cell is also calculated using a difference equation. The common characteristic in these models is that speed is usually a function of cell occupants, upstream cells speed, and downstream cells density [38–40].

3.8 Traffic Assignment Models

The goal of traffic assignment is to model route choice and congestion in urban areas mainly during peak periods, which facilitates evaluation transportation infrastructure performance [41]. In these models, demand between each origin and destination in a network is estimated and assigned to arcs in the network. Models of traffic flow theory is used here to determine arcs travel time and cost. The importance of these models is that they allow travel time to be a function of flow (or density). For the purpose of this book, only a simple approach for solving such models is presented follows for illustration [42]. In this approach, traffic demands are assigned sequentially and iteratively to the network. An origin–destination pair is selected and its demand is assigned usually to the shortest path between the origin and destination up to the available capacity. Then network capacities are updated and travel times are recalculated until all traffic demand is satisfied. Lu et al. proposed a heuristic for evacuation routing that takes the idea of traffic assignment even though it assumes that travel time is not a function of flow [43]. Their model, called capacity constrained route planner (CCRP), finds evacuation routes and schedules that minimize total evacuation time while allowing time-dependent node and arc capacities. The heuristic iteratively allocate evacuees to routes with the earliest arrival time to safety from any source node. In each iteration, the algorithm searches for the earliest arrival route to safety, allocate evacuees to routes, and update capacity and remaining evacuees. The heuristic contuse until all evacuees reach safety.

Chapter 4
Integrated Model for Evacuation Planning

Abstract This chapter presents an integrated model for evacuation planning. In this methodology, an optimization algorithm is integrated into a traffic simulation routine in order to determine the best routes for evacuation. The main drawback in simulation models is that they do not have the capability of identifying the "*best*" routes (optimization) and therefore, a model that integrates an optimization and a simulation routine into one algorithm is proposed and used for evacuation route planning. It also includes examples to illustrate the application of the implemented methodology.

4.1 Introduction

The problem addressed in this book is the evacuation problem, which is stated as, "*Given a specified number of evacuees and their locations, what are the best evacuation routes and schedule to evacuate all people within minimum amount of time.*" The same problem also has similar applications in logistics and communication networks. In evacuation planning, the transportation infrastructure is modeled by a network where intersections are represented by nodes and roads and highways by arcs. The following simple example illustrates the problem and network modeling. In this example, 100 vehicles (or units of flow) have to be evacuated from node 1 to node 6 with minimum amount of time. First let us assume that travel time (τ_{ij}) and capacity (u_{ij}) on each road (arc) are constant as given in Fig. 4.1.

This example can be solved by inspection. Only one path flow can leave the source node at any time (because only one arc is leaving the source). Therefore, the goal is to select the best path to send the flow at each time period. At time "0," the path 1-2-5-6 is the shortest path and has a capacity of 10 (it is the minimum capacity on all the arcs that belong to this path). Therefore, we can send a flow of 10 along this path. Actually, since travel time and capacity are constants, the same path will be always favorable over the other two alternative paths (1-2-3-5-6 and 1-2-4-5-6). Consequently, the minimum evacuation time of 15 time units can be achieved by

A. Naser and A.K. Kamrani, *Intelligent Transportation and Evacuation Planning:*
A Modeling-Based Approach, DOI 10.1007/978-1-4614-2143-6_4,
© Springer Science+Business Media, LLC 2012

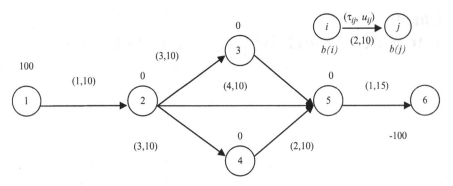

Fig. 4.1 Network representation for Example 5.1

sending a flow of 10 along path 1-2-5-6 ten times starting at time "0" until time "9." The last flow leaving node 1 at time 9 will arrive at node 6 at time 15.

In this example, ten chains of flow are repeatedly leaving the source node 1 and heading toward safety at node 6. Any model assuming constant travel time will continually choose path 1-2-5-6 for each flow leaving node 1. To demonstrate the problem of using constant travel time assumption, let us assume that a flow entering an empty arc will travel through the arc at free flow speed (maximum speed for that arc). If a flow enters an occupied arc, the flow will travel through the arc at a speed defined by the Greensheild equation (3.1). Let us assume that arc (5,2) has the following parameters: distance is 4 miles, free speed is 1 mile/h, and jam density 20 veh./mile. Now, let us assume that the fifth flow chain leaving the source arrives at node 2 and there are already 30 vehicles on arc (2,5). Using (3.1), the expected speed through that arc drops to 0.625 miles/h and, thus, a travel time of 6.4 h. With this travel time, using arc (2,5) is no longer favorable over using the other two alternatives (2-4-5 and 2-3-5), which have a travel time of 5 h.

This result shows that constant travel time assumption can lead to inaccurate and ineffective result. A study of previous work on this topic indicates that analytical models are efficient in finding the best routes for evacuation, and simulation models are efficient in predicting and estimating travel time as a function of flow. Using this observation, the author developed a new model that integrates an optimization routine and a traffic simulation model into one algorithm. In every iteration of this algorithm, an optimization routine directs the flow leaving an arc, and a simulation routine propagates the flow along the selected arcs until all evacuees reach safety. Figure 4.2 shows a general structure of the new model.

4.2   General Structure of the Integrated Model

Simulation models, in general, are based on discretizing the time space and updating system variables at each time step. This time discretizing facilitates capturing nonlinear and usually complex system characteristics. To link a

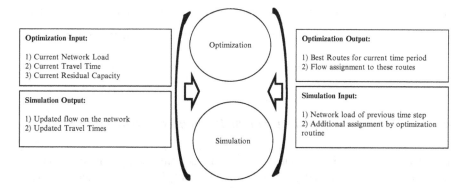

Fig. 4.2 Diagram for optimization and simulation routines

simulation model to an optimization routine, naturally, we will try to discretize the time space for the optimization model as well. Consequently, the proposed general structure for the new router is to divide the time space into smaller and equal time intervals.

At each time step, an instance of the problem is solved through the optimization model, and system variables are updated through the simulation model. A major benefit for discretizing the time space in the evacuation problem is to facilitate capturing the nonlinearity of load-dependent travel time. This approach is much more favorable than the other alternative of building a very complex analytical model that most likely will have no efficient solution technique. At each time step of the proposed model, travel time and arcs capacities are updated based on current arcs load. Based on these updated parameters, the optimization routine selects the best routes to direct the flow out of each supply node. Once this flow is assigned to an arc, the simulation routine propagates and move the flow along that arc until it reaches the head node. The algorithm terminates when all evacuees reach safety. The following is the general framework of the algorithm.

Begin

$T = 0$

While there are still evacuees not reaching safety

(*optimization Routine*)

Estimate arcs travel times and capacity base on current load
Load traffic to arcs that maximize flow to safety (augment flow) for the current period

(*Traffic Propagation Routine*)

Simulate the flow along each arc for one time period using any appropriate traffic simulation and prediction model

$T = T + 1$

End while
Minimum evacuation time = T

End

In order to fully describe the new model, two questions have yet to be answered:

- How can we maintain an equivalent problem to the original network at each time step?
- How can we link the instance of the problem at time T to the one at time $T + 1$?

4.2.1 Maintaining an Equivalent Problem at Each Time Step

In order to maintain an equivalent problem at each time step, total supply in the network has to be conserved. In other words, no supply is allowed to vanish nor is it allowed to be created. The following relation is used to conserve the supply at any time step T:

- Total supply in the original network = current supply at time T + flow traveling the arcs + total flow reached safety by time T.
- The supply at each node is updated. If a flow leaves a node, the supply is decreased, and if a flow enters the node the supply is increased.
- The total flow reached safety is updated. All flow leaving the arcs leading to the sink node is added to flow reaching safety.
- Total flow in any arc = total flow at time $T - 1$ + incoming flow − leaving flow.

When finding the best routes for current time period, the optimization routine needs the following updated parameters: travel time, arcs capacity, and nodes supply. The problem is how to account for the flow already on the arcs to maintain an equivalent instance of the original network. The idea used here is that this flow will arrive at the head node at a future time. Recalling in previous section, a dynamic network optimization can be solved using a static algorithm if the original network is transformed into a time-expanded network. In the time-expanded network, each node has a copy in each future time period. Therefore, when constructing the time-expanded network to solve an instance of the problem at time T, the flow on the arcs can be assigned as a supply to future copies of the head node. For example, assume there is a flow of 10 traveling through an arc (i,j), and this flow is expected to arrive at node j in four time periods. Then, when constructing the time-expanded network that contains arc (i,j), a supply of 10 is assigned to the copy of node j at time 4.

4.2.2 The Link of an Instance of the Problem
at Time T to the One at Time $T + 1$

The proposed model divides the time space into smaller and equal time intervals. At each time step, the optimization routine solves an instance of the problem and

Fig. 4.3 Network representation

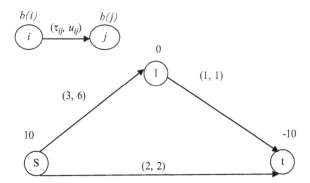

the simulation model updates system variables. It may be easy to link, in the simulation model, an instance of the problem at time T to the one at time $T + 1$. However, for the optimization routine this linkage needs further analysis. Jarvis and Ratliff, using Theorem 1 below, showed that solving the minimum cost turnstile problem is equivalent to solving other dynamic objectives [30]. These objectives include:

- The number of evacuees existing in periods 1 through t is maximized for all values of t.
- The average number of periods an evacuee required to exist is minimized.
- The period in which the last evacuee exists is minimized.

Theorem 1 [30] Consider any feasible flow with a value v from s to t within time horizon T and the following three objectives:

(a) Maximize $v(P)$ for $P = 1, 2, \ldots, T$

(b) Minimize $\sum_{t=0}^{T} c_t(v(t) - v(t - 1))$ where $v(-1) = 0$ and $c_1 < c_2 < \ldots < c_T$

(c) Minimize P such that $v(k + P) = v(P)$ for $k = 1, 2, \ldots, T - P$

Any feasible flow of v from s to t that satisfies either objective (a) or (b) will also satisfies the other two. The part of interest in this theorem is that solving an earliest arrival flow problem with a maximum flow value of v is equivalent to solving the minimum evacuation time (quickest transshipment) problem with an initial supply equals to v. By definition, a solution to the earliest arrival problem up to time T is also a solution to the problem up to time $T - 1$. In other words, we can start with the current earliest arrival flow at time T to construct the earliest arrival flow at time $T + 1$. Therefore, a key to link an instance of the problem at any time T to the one at time $T + 1$, is to maintain (or always to solve for) the earliest arrival flow at each time step. A drawback for this theory is the restriction to the evacuation problems where only one sink (safety node) exists. The reason is that the earliest arrival flow doses not necessarily exist in a problem with multiple sinks. In the following example, ten evacuees are located at node S and the goal is to send them to node t with the minimum amount of time (Fig. 4.3).

Table 4.1 Input parameters

Arc	Distance, D (miles)	Free flow speed, v_f (miles/h)	Jam density, k_j (veh./mile)	Capacity, U
$(S,1)$	3	1	12	6
(S,t)	2	1	4	2
$(1,t)$	1	1	2	1

Fig. 4.4 Cell representation

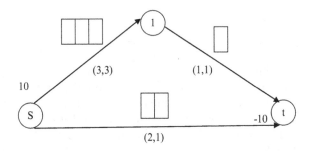

Table 4.2 Optimization routine input

Node	$B(I)$	Arc	$T_{ij}(T)$	$U_{ij}(T)$
1	10	$(S,1)$	3	2
2	0	(S,t)	2	1
3	-10	$(1,t)$	1	1

The parameters given in Table 4.1 are assigned to the network.

The purpose of this example is to show the general flow and progress of the proposed methodology. The optimization routine is presented in Chap. 5; therefore, it is presented in this example as a black box, where inputs and outputs only are explained. The simulation routine, in this example, is very simple flow propagation model. Each arc is divided into a number of cells that equals to D/v_f (Fig. 4.4). At each time step t, the simulation routine will update the number of evacuees exist at each cell k, $n_{ijk}(t)$. This simple model assumes that all evacuees in a cell k will advance to cell $k + 1$ when the clock advances from t to $t + 1$. This is basically the case when travel time is constant (i.e., vehicles always travel at free flow speed). In more realist settings, the speed is not constant and it is a function of current occupants in the cells. A network at $T = 0$ is illustrated in Fig. 4.4.

4.2.3 Optimization Routine

Input. Current supply at each node and updated arcs travel time and residual capacity (Table 4.2) and updated supply $b_i(t) = b_i(t - 1) + \text{inflow} - \text{outflow}$.

Table 4.3 Optimization output

Arc	$X_{ij}(T)$
(S,1)	1
(S,t)	1
(1,t)	0

Table 4.4 Cell occupancies from last time period and the output of the optimization

Arc	$X_{ij}(T)$	$N_1(T)$	$N_2(T)$	$N_3(T)$
(S,1)	1	0	0	0
(S,t)	1	0	0	0
(1,t)	0	0	0	0

Table 4.5 Updated cell occupancies

Arc	$N_1(T+1)$	$N_2(T+1)$	$N_3(T+1)$
(S,1)	1	0	0
(S,t)	1	0	0
(1,t)	0	0	0

Table 4.6 Evacuation plan

Number of evacuees	Depart time	Path	Time of arrival at safety
1	0	S-1-t	4
1	0	S-t	2
1	1	S-1-t	5
1	1	S-t	3
1	2	S-1-t	6
1	2	S-t	4
1	3	S-1-t	6
1	3	S-t	5
1	4	S-t	6
1	5	S-t	7

Output. Routes and flow for the current period t, $X_{ij}(t)$. This flow will enter the first cell for arc (i,j), i.e., $n_1^{(i,j)}(t) = X_{ij}(t)$. Let us assume in this example, and based on the current nodes supply and travel times, the output is as shown in Table 4.3.

4.2.4   Simulation Routine

This step will update the number of evacuees that exist at each arc cell (Tables 4.4–4.6 and Fig. 4.5).

Output. The algorithm terminates with total evacuation time = 7.

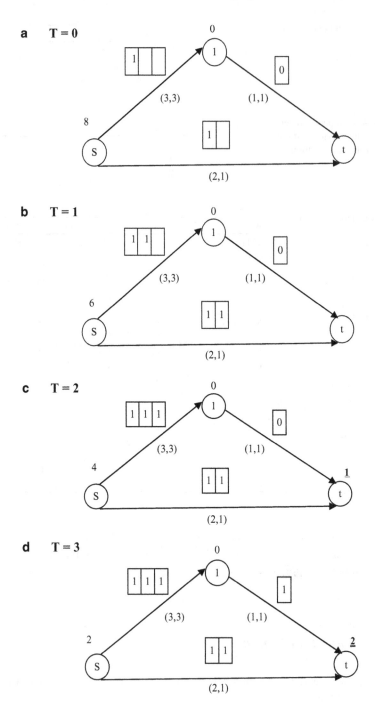

Fig. 4.5 Algorithm iterations

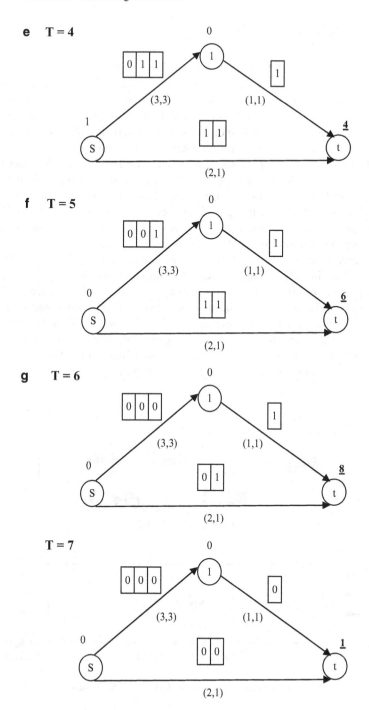

Fig. 4.5 (continued)

4.3   Traffic Simulation Model

Traffic simulation models, in general, can be classified into microscopic and macroscopic models. The classification is based on the required level of resolution (or abstraction) for traffic operations. Whereas, individual vehicles and driver's behavior are captured in microscopic models, these entities are aggregated in macroscopic models into a homogenous flow. Examples of traffic operations modeled and traced in microscopic simulation include car following patterns and individual driver's decision of acceleration and deceleration. On the other hand, most macroscopic simulations model traffic as a homogenous fluid and trace aggregate variables such as average speed, flow rate, and density (Fig. 4.6). One can expect that even though microscopic modeling captures more details and can produce more accurate results, it is ineffective for large-scale networks. This is due to the complexity and the size of computations involved.

A macroscopic model is adopted in this model for two reasons [50, 52]. In addition to simplifying computations, the output of the simulation is linked to an optimization routine. Most analytical evacuation models in literature solve the macroscopic version of the problem where flow is homogenous and individual behavior is ignored. Using macroscopic modeling for the simulation facilitates this linkage, whereas a transformation and aggregation of variables have to be done if the simulation is microscopic. In macroscopic modeling, traffic state is defined by three variables: flow rate, average speed, and density (Fig. 4.7).

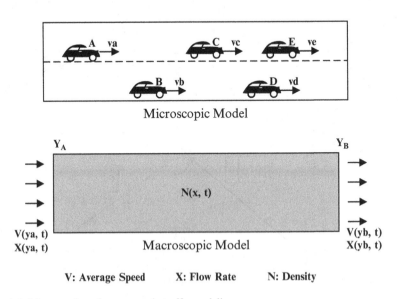

Fig. 4.6 Microscopic and macroscopic traffic modeling

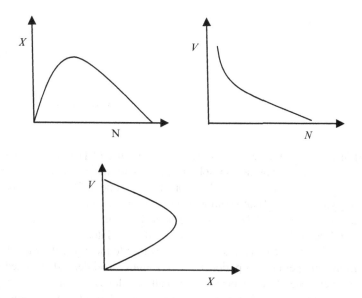

Fig. 4.7 Relationship among traffic variables

- Flow rate (X) is number of vehicles passing through a point per unit of time (veh./h).
- Average speed (V) is the average distance traveled per unit time (miles/h).
- Density (N) is the number of vehicles per unit distance (veh./mile).

The fundamental equation linking traffic variables is shown below:

$$X = VN. \tag{4.1}$$

The following general graphs (Fig. 4.7) are common in traffic theory literature describing the relation among these variables [47].

One of the earliest macroscopic models found in the literature is due to Lighthill and Witham [35]. Let $N(y,t)$ and $X(y,t)$ be the density and flow rate of traffic at location y and time t, respectively. Considering a length of road dy and time interval dt, the number of vehicles on dy at time t is $N(y,t)dy$ and the number of vehicles entering location y during dt is $X(y,t)dt$. Conservation of vehicles implies that

$$[N(x,t+dt) - N(x,t)]dx = [Y(x,t) - Y(x+dx,t)]dt$$

or

$$\frac{\partial N(y,t)}{\partial t} + \frac{\partial X(y,t)}{\partial y} = 0. \tag{4.2}$$

Fig. 4.8 A road divided into homogenous cells

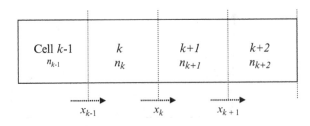

Most macroscopic traffic models in the literature use (4.1) and (4.2) to simulate traffic. The common approach in solving (4.2) is to solve a series of difference equations by discretizing time and location spaces. Time and space are discretized by dividing each arc (road) into homogenous cells and dividing time into smaller and equal time intervals (Fig. 4.8). The length of each cell is usually set at the distance traveled at free flow speed during the selected time interval. Under light traffic conditions, vehicles in a cell are assumed to advance to the next cell when the clock moves from time period t to period $t + 1$. Assuming load-dependent speed (travel time), however, congestion can occur, and not all vehicles can advance to the next cell.

Several models are available in the literature, which use the cell representation of road segments [39, 41–43]. These models differ mainly in average speed equation. Earlier models define the system state at each time interval as the number of vehicles exists at each cell k on arc (i,j), $n_{ijk}(t)$, according to the following recursive equations:

$$n_{ijk}(t + 1) = n_{ijk}(t) + \frac{l_{ijk}}{\Delta t}[x_{ijk-1}(t) - -x_{ijk}(t)], \qquad (4.3a)$$

where l_{ijk} is the length of cell k on arc (i,j) and Δt is the time interval. Equation (4.3a) basically means that the number of vehicles in cell k at time $t + 1$ equals to the number of vehicles existed at time t plus incoming vehicles minus leaving vehicles. Flow rate is estimated using the fundamental equation (4.1)

$$x_{ijk}(t) = v_{ijk}(t)n_{ijk}(t) \qquad (4.3b)$$

and it refers to the number of vehicles that passes from cell k to cell $k + 1$ during the interval $[t, t + \Delta t]$. The average speed is calculated as a function of density:

$$v_{ijk}(t) = f(n_{ijk}(t)). \qquad (4.3c)$$

Equation (3.3) completely defines the system at each time step. Speed–density models ($V = f(N)$) are rich research topic in traffic literature. These models are usually developed through fitting real speed and density data with an empirical equation. A famous linear model was done by Greenshield [19]:

$$v = v_f\left(1 - \frac{n}{n_j}\right), \qquad (4.4)$$

where v_f is the free flow speed and n_j is the jam density. Another model was due to Greenberg as follows [48]:

$$v = c \, \ln\left(\frac{n}{n_j}\right),$$
(4.5)

where c is a constant. In general, speed–density functions in the literature have the following form [43]:

$$v = v_f\left(1 - \left(\frac{n}{n_j}\right)^l\right)^m, \quad l > 0, \quad m > 1.$$
(4.6)

Models that have only density as the only difference equation (called first-order macroscopic) showed poor transient behavior [41]. More sophisticated second-order models have been developed to address shortcoming of previous models [41–43, 49]. In these models, average speed is also calculated through a difference equation as follows:

$$v_{ijk}(t+1) = v_{ijk}(t) + \frac{\Delta t}{\tau}[V_e(n_{ijk}) - v_{ijk}(t)] + \frac{\Delta t}{l_{ijk}}v_{ijk}(t)[v_{ijk-1}(t) - v_{ijk}(t)]$$

$$- \frac{\mu \Delta t}{\tau l_{ijk}}\left[\frac{n_{ijk+1}(t) - n_{ijk}(t)}{n_{ijk}(t) + \kappa}\right].$$
(4.7)

τ, μ, κ are constants to be estimated through calibration, $V_e(n_{ijk}(t))$ is the equilibrium speed found through a speed–density equation like (4.6). Several notes can be mentioned about the speed model (4.7):

- The second term in (4.7) is called the relaxation factor. It expresses that the mean speed tends asymptotically toward the equilibrium speed defined by the fundamental speed–density diagram. "τ" determines the rate at which speed should move toward equilibrium.
- The third term is the convection factor. It represents the influence of incoming traffic speed.
- The last term is the anticipation factor. It considers the fact that the average speed in cell k is affected by upstream density. τ, μ, κ affect the rate of this factor.

Proposed simulation model is as follows:

Δt be the selected time interval
d_{ij} be the distance of arc (i,j)
l_{ijk} be the length of cell k on arc (i,j)
v_{fij} be the free flow speed of arc (i,j)
U_{ij} be the maximum flow rate (capacity) of arc (i,j)

Fig. 4.9 Dividing each arc into homogenous cells

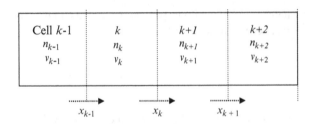

u_{ijk} be the maximum flow rate leaving cell k (entering cell $k + 1$) on arc (i,j)
N_{ij} be the jam density of arc (i,j)
N_{ijk} be the jam density of cell k on arc (i,j)
$n_{ijk}(t)$ be the number of vehicles traveling in cell k on arc (i,j) at time t
$x_{ijk}(t)$ be the number of vehicles leaving cell k on arc (i,j) into cell $k + 1$ traveling
               in cell k when the clock advances to $t + \Delta t$
$v_{ijk}(t)$ be the average speed of vehicles traveling in cell k on arc (i,j) at time t

Each road is divided into homogenous cells (Fig. 4.9) with the following preliminary calculations:

The largest Δt that can be selected $= \min_{ij \forall (i,j) \in A}(d_{ij}/v_{f\,ij})$
For each arc (i,j) the length of each cell k is $l_{ijk} = v_{f\,ij} \times \Delta t$
The number of cells in arc$(i,j) = (d_{ij}/l_{ij})$
Cell's capacity $u_{ijk} = U_{ij} \times \Delta t$
Cell's jam density $N_{ijk} = N_{ij} \times l_{ijk}$

The model updates the number of vehicles in each cell and their average speed according to the following set of equations:

$$n_{ijk}(t+1) = n_{ijk}(t) + \frac{l_{ijk}}{\Delta t}[x_{ijk-1}(t) - x_{ijk}(t)], \tag{4.8a}$$

$$v_{ijk}(t+1) = v_{ijk}(t) + \frac{\Delta t}{\tau}[V_e(n_{ijk}) - v_{ijk}(t)] + \frac{\Delta t}{l_{ijk}}v_{ijk}(t)[v_{ijk-1}(t)$$
$$- v_{ijk}(t)] - \frac{\mu\Delta t}{\tau l_{ijk}}\left[\frac{n_{ijk+1}(t) - n_{ijk}(t)}{n_{ijk}(t) + \kappa}\right], \tag{4.8b}$$

$$x_{ijk}(t) = v_{ijk}(t)n_{ijk}(t), \tag{4.8c}$$

$$V_e(n_{ijk}) = v_f\left(1 - \left(\frac{n_{ijk}}{N_{ijk}}\right)^l\right)^m, \quad l > 0, \ m > 1, \tag{4.8d}$$

$$n_{ijk}(t+1) \leq N_{ijk}$$

Fig. 4.10 Boundary
conditions for arc (i,j)

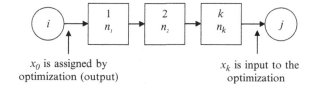

x_0 is assigned by x_k is input to the
optimization (output)   optimization

or

$$x_{ijk}(t) \leq \frac{\Delta t}{l_{ijk+1}} \left[N_{ijk+1}(t) - n_{ijk+1}(t) \right] + x_{ijk+1}(t), \qquad (4.8e)$$

$$x_{ijk}(t) \leq u_{ijk}. \qquad (4.8f)$$

4.3.1   Boundary Conditions

Assuming infinite capacity at the head node, all flow leaving the last cell will be
assigned as a supply to that node. The optimization routine determines the "best"
routes and flow assignment to direct the flow out of the supply (head) node
(Fig. 4.10).

4.4   The Optimization Routine

Several analytical models have been developed in the literature to solve the
evacuation problem. The common characteristic in these models is utilizing
dynamic network optimization algorithms and their effective solution
methodologies. Most of these models, however, assume constant travel time.
To overcome this challenge, the proposed model updates travel times at each
time step as a function of flow, and applies an optimization routine to the updated
problem. This section presents the optimization routine in details.

The evacuation problem is similar to the quickest transshipment problem (QTP)
(Fig. 4.11) available in the literature. In this problem, the goal is to find the
minimum time to move a positive supply (evacuees) from supply (danger) node
(s) to a sink (safety) node. The special case of this problem is the quickest flow
problem with one source and one sink node. Let us first study a method for solving
the quickest transshipment problem. The quickest transshipment problem can be
represented by graph $G(N,A,B)$ where N is the set of nodes, A is the set of arcs, and
B is a supply vector. A feasible solution of the problem is to find a feasible instance
$G(N,A,B,T)$ where T is the time horizon (QTP^T). An optimum solution is a feasible
instance with the minimum time horizon possible.

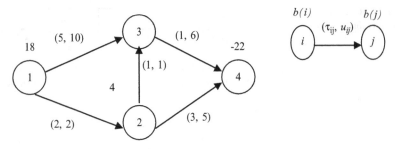

Fig. 4.11 An example of QTP. Source nodes are 1, 2. Sink node is 4

A favorable property for the QTP is usually sought, and evacuation application is the earliest arrival property. A QTP with earliest arrival flow, not only satisfy the supply within minimum amount of time, it guarantees also that the amount of flow reaching sink is as much as possible at any smaller time within the minimum time horizon.

An approach for solving QTP is to perform a binary search for the minimum time horizon and then to find a feasible flow within the minimum time. The search begins in the middle of a time interval between known lower and upper bounds for the minimum time. When we select a time T at the middle of the time interval we have an instance $G(N,A,B,T)$ of the quickest transshipment problem. The next step is to test the feasibility of this instance (below is more details on testing feasibility). If the problem is infeasible then the lower bound is updated to equal T, and if it is feasible the upper bound is updated to T. Consequently, after each search and feasibility testing, the solution space (time interval between upper and lower bounds) becomes smaller either by increasing the lower bound or decreasing the upper bound. The algorithm stops when the gap between the upper bound and the lower bound is less than a predetermined value (minimum gap is one). The following is the general structure of the algorithm:

Begin
        Find *UB, LB,* ε;
        While $(UB - LB) > \varepsilon$
            $T = (UB + LB) / 2$
            Test feasibility of $G(N,A,B,T)$
            if feasible
                $UB = T$;
            Else
                $LB = T$;
            End if
        End while
        Minimum time = UB
End

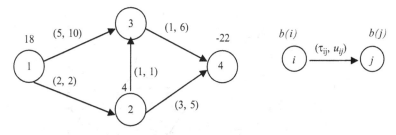

Fig. 4.12 Network representation

To determine the feasibility of any instance of the problem, and to find that feasible flow, a super source is added to the network with arcs connecting to each original source. The capacity of each added arc equals the supply at these sources. A maximum flow problem is then solved from the super source to the sink. The problem is feasible if, and only if, all arcs connecting the super source are saturated (flow equals capacity). In aiming for the *earliest arrival flow*, the algorithm augments along the first available shortest distance path from super source to sink. The algorithm is outlined below:

Begin

        Generate Time expanded network for $G(N,A,B,T)$;

        Add super source S;

        Add arc (S,i) with $u_{si} = b(i)$ $\forall i$ where $b(i) > 0$;

        Solve shortest path from S to sink using τ_{ij} as distance

        While $d[\text{sink}] < (T + 1)$ " *while shortest path to sink is within time horizon T"*

        Begin

                Find the shortest path from S   to sink. Let δ be the min residual capacity;

                Augment the flow on the path with δ;

                Solve shortest path from S to sink on the residual network;

        End While

        IF $x_{Si} = u_{si} \ \forall i$ where $b(i) > 0$ Problem is feasible. Otherwise not feasible;

End

Consider the network in Fig. 4.12. The numbers on each arc represent travel time and capacity, respectively. Numbers above nodes represent supply (demand when negative). The goal is to send the supply at nodes 1 and 2 to the sink (node 4) within the minimum time.

The solution is described as follows:

1. *Find lower and upper bounds*: An easy lower bound is 0. However, having a tight interval between the lower and upper bounds always speeds the algorithm.

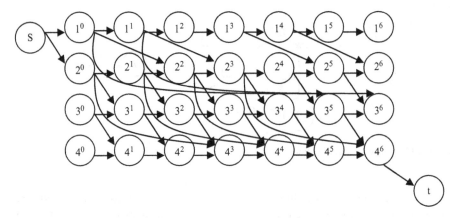

Fig. 4.13 Time-expanded network

Therefore, we will try to find a tighter lower bound as shown below. Considering travel time as a distance, the shortest path from the source to the sink is either a lower bound or the optimum solution. This is an immediate indication that no flow can reach the sink faster than the shortest path, and most of the time, the supply at the source exceeds the capacity of the shortest path. In the above example, the shortest path is S-2-3-4 with time of 2. Since the capacity of this path is 1 (it is the minimum capacity among all arcs on the path) LB can be then set at 2. To find an upper bound, we aim at finding "arc-disjoint" paths from the source to the sink. These paths will have no arc in common. In the above example, two arc-disjoint paths are S-1-3-4 with a capacity of 1 and S-2-4 with a capacity of 3. As a worst case scenario the supply at node 1 can be satisfied by sending several "shipment" along path S-1-3-4 and the same is true for node 2 along path S-2-4.

To find the time needed to send supply at 1 along path S-1-3-4,

$$T_1 = \text{length of path} + \left\lceil \frac{\text{supply at node 1}}{\text{path capacity}} \right\rceil$$

$$= 6 + \left\lceil \frac{18}{6} \right\rceil$$

$$= 9$$

Using the same procedure, $T_2 = 3$ and therefore UB $= \max \{9, 4\} = 9$.

2. A binary search can now start in the interval [2,9] to find the minimum transshipment time with a required gap of 1, as follows:

(a) $T = 2 + \lceil (9 - 2)/2 \rceil = 6$.
(b) Check if the problem is feasible at $T = 6$. First construct the time-expanded network (Fig. 4.13).

Table 4.7 Solution

Path	Flow	Depart time	Arrive at sink
2-3-4	1	0	2
2-4	3	0	3
1-2-3-4	1	0	4
1-2-4	1	0	5
1-2-3-4	1	1	5
1-2-4	1	1	6
1-3-4	6	0	6
1-3-4	6	1	7
1-2-4	2	2	7

Fig. 4.14 Cell occupants as "future" supply

(c) The maximum flow solution is 14 which does not saturates the arcs is $(S,1^0)$ and $(S,2^0)$. Therefore, the problem is infeasible.

(d) Set LB = 6.

$$\lceil T \rceil = 6 + \frac{(9-6)}{2} = 8.$$

(e) Check if the problem is feasible at $T = 8$. The maximum flow solution is 23; therefore, it is feasible. Set UB = 8, and $T = 7$.

(f) Check if problem is feasible at $T = 7$. The maximum flow solution is 23; therefore, it is feasible. Set UB = 7.

(g) Since UB − LB ≤ 1 the algorithm terminates with minimum time = 7.

(h) To construct an earliest arrival flow within the minimum time, the procedure selects the paths with the minimum travel time first as given in Table 4.7.

At each iteration of the proposed methodology, an earliest arrival quickest transshipment problem is solved. Network parameters including the supply vector, arcs capacities, and travel times are updated and used as input to the optimization routine. However, several modifications are proposed in this methodology when solving the QTP. In order to maintain an equivalent problem the occupants at each cell have to be considered when updating the supply vector. In Fig. 4.14, flow leaving cell 3 is immediate supply to node j or copy "0" in the time-expanded network. However, occupants left at cells 1–3 are not expected to arrive at node j until later times. Therefore, they are considered future supply to node j. The time-expanded structure facilitates this assignment as we have a copy for each node at each future time period.

A major benefit for the proposed model is the ability to consider load-dependent travel times. At each time step, travel time and the supply vector are updated. The new model takes few virtual time steps in the future to estimate when current cell occupants will arrive at the head nodes and also to estimate how long it takes to

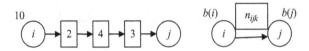

Fig. 4.15 Example of updating the supply vector

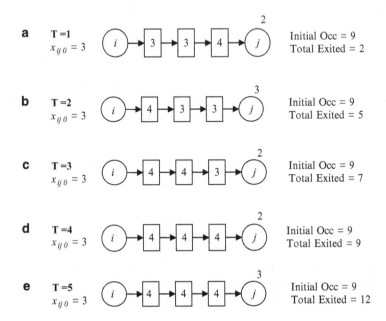

Fig. 4.16 (a–e) Example for subsimulation and updating the supply vector

travel the arc. It is like if the model freezes the actual clock temporary and run a "subsimulation" and then goes back to previous settings when it is done. Although, there are other alternatives to estimate travel time and future nodes capacity that require less computations, this "subsimulation" showed superior results. The subsimulation uses the same simulation model and the same equations for flow rate and average speed. It terminates when all arcs are emptied from current occupants. At each time step T in the virtual clock, the amount of flow leaving the arc is assigned as a supply to the head at future copy "T" (recall that in the time-expanded network each node has a copy at each future time period). As soon the last flow leaves the arc, the model record the time in the virtual clock as the travel time for the arc. For example, consider the following arc (Fig. 4.15).

Assuming the $x_{ij0} = 3$, $x_{ij3} = 2$. Then supply at tail and head nodes is updated as follows: $b(i) = 10 - 3 = 7$ and $b(j) = 0 + 2 = 2$.

Occupants in cells 1–3 has still to be considered. These occupants are supply to node j but at future times. Assuming arc's capacity equals 3, and assuming the following subsimulation is performed (no details here on speed and flow calculations) (Fig. 4.16).

Table 4.8 Supply
assignment for node j

Copy of node, j	Supply
1	2
2	3
3	2
4	2

At this point, the subsimulation terminates with the results presented in Table 4.8.

Travel time is estimated at 5. Travel time can be looked at as the time it takes for a new flow into the arc (the first x_{ij0} at $T = 1$) to exit the arc. The arc emptied the initial nine occupants at time 4 and the new flow of three exited at time 5. After the subsimulation terminates the arc, it then returns to the previous setting as in Fig. 4.16. At each step, arcs capacities (U_{ij}) are updated to reflect the number of vehicles existed in the first cell. The rule here is to prevent the number of vehicles in the first cell from exceeding cell's jam density. That is

$$U_{ij}(t) \leq [N_{ijk} - n_{ijk}]. \tag{4.9}$$

4.5   The Integrated Model

The proposed model is constructed, briefly, by integrating an optimization routine into a simulation routine at successive time periods. At each time step, the optimization routine finds the optimum routes that minimize total evacuation time, and the simulation routine "partially" propagates traffic along these routes while satisfying main traffic theories. During the algorithm, traffic is directed toward safety regions and the total number of evacuees reaching safety is updated at each time step. When the number of evacuees reaching safety equals the initial total number of evacuees the algorithm terminates. Below are the general algorithm and its flow diagram (Fig. 4.17).

Begin

   $T = 0$

   While there are evacuees not reaching safety

      Update arcs travel times and node supply

      (*optimization Routine*)

      Load traffic to arcs that minimizes Total Evacuation Time

      (*Traffic Propagation Routine*)

      Simulate the flow along each arc for one time period

      $T = T + 1$

   End while

   Minimum evacuation time = T

End

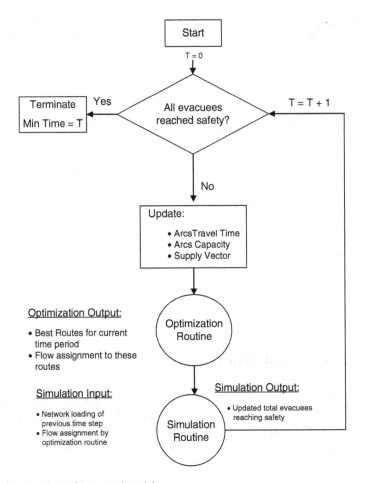

Fig. 4.17 Flow chart of integrated model

Let N be the set of nodes and A be the set of arcs; the formal algorithm with exact calculations is followed:

Step I:

1. Perform the preliminary calculations
2. let $t = 0$;

Step II (Optimization):

Given current $b(i)$, $\tau_{ij}(t)$, $U_{ij}(t)$ *solve the quickest transshipment problem*
Output is flow: $x_{0ij}(t) \; \forall \; (i,j) \in A$

Step III (Simulation):

1. Update $n_{ijk}(t) \ \forall \ (i,j) \in A, k = 0, 1, \ldots, c_{ij}$ as in (4.8a)
2. Update $v_{ijk}(t) \ \forall \ (i,j) \in A, k = 0, 1, \ldots, c_{ij}$ as in (4.8b) and (4.8d)
3. Find $x_{ijk}(t) \ \forall \ (i,j) \in A, k = 1, 2, \ldots, c_{ij}$ as in (4.8c), (4.8e), and (4.8f)
4. Update travel times $\tau_{ij}(t)$ and supply $b(i)$ using the subsimulation procedure
5. Update capacity $U_{ij}(t)$ as in (4.9)

Step IV:

If $\sum_{i \in N} |b(i)| = 0$, STOP minimum time equal t.
Otherwise $t = t + 1$; go to Step II

Chapter 5
Model Evaluation and Experimentation

Abstract The purpose of this chapter is to test the premises of the presented methodology and its capability to plan an optimum path for evacuation. Several sample cases using the proposed methodology are presented. To achieve this goal, a traffic simulator is developed that evaluates the optimum solution obtained by a constant travel time algorithm.

5.1 Introduction

The previous chapter presented a new methodology for solving a dynamic routing problem with a special application in evacuation and traffic routing. The new methodology is intended to improve the solution produced by previous analytical models that assume constant travel time. The main premise of this methodology is that the quality of even the optimum solutions of these previous algorithms, may suffer significantly when applied in more practical settings. More specifically, let us assume that the optimum solutions of a constant travel time algorithm are simulated through a traffic model that mimics real life conditions. It is claimed here that many of these solutions may cause congestion and therefore may produce poor results. The presented methodology is hoped to produce more superior results that outperform the poor results of constant travel algorithms.

The purpose of this chapter is to test the premise of the presented methodology. To achieve this goal, a traffic simulator is developed that evaluates the optimum solution obtained by a constant travel time algorithm. The application considered in this research is evacuation planning, and therefore, the selected constant travel time algorithm is for the quickest transshipment problem. Briefly, the simulator first solves the quickest transshipment problem with constant travel times. Then, the solution obtained is decomposed into chains (paths) from source (danger nodes) to sink (safety node). The resulted chains are then simulated and the actual evacuation

A. Naser and A.K. Kamrani, *Intelligent Transportation and Evacuation Planning:*
A Modeling-Based Approach, DOI 10.1007/978-1-4614-2143-6_5,
© Springer Science+Business Media, LLC 2012

time is recorded. For the same problem, a solution is obtained by using the proposed new methodology and the two evacuation times are compared. The remaining of this chapter is organized as follows. Section 5.2 presents the simulator and discusses flow decomposition. The following section presents several network case studies taken from different parts of Houston and Galveston. The structure of the experiment is also presented in this section. Section 5.4 presents the numerical results; and finally, the last section presents a discussion and a summary.

5.2   Flow Decomposition

Chapter 4 presented a binary search methodology for solving the quickest transshipment problem. In order to simulate a solution for this problem, it is necessary to decompose the solution into chain paths from sources to the sink. The solution is originally represented by the amount of flow on each arc at the beginning of each time period. However, to simulate such a flow, it is necessary to determine the route this flow will follow to reach the sink. For example, consider the flow on the network shown below. It is easy to determine from the data shown that the amount of flow reaching safety is 7. However, it is not exactly clear how this flow is traveling from the source to sink (Fig. 5.1).

Decomposition for the above flow can be represented as follows:

Path 1-3-2-4 with a flow of 1
Path 1-3-4 with a flow of 3
Path 1-2-3-4 with a flow of 2
Path 1-2-4 with a flow of 1

An algorithm for flow decomposition is presented next [16]. The algorithm is a path search routine that uses only the arcs on the network that has a positive capacity. The *capacity of each arc is set initially to equal its flow.* A search first finds a path from a source to the sink. The algorithm then finds the arc with the

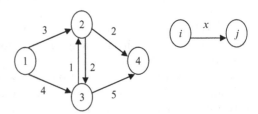

Fig. 5.1 Example of flow decomposition

minimum flow and augments this flow from the path. The result of each path search and augmentation is that at least one arc is deleted from the network (one with the minimum capacity). The algorithm continues on searching for paths from source to sink until no further paths exist. The following is a general algorithm:

Flow Decomposition Algorithm

Begin

>Let $u_{ij} = x_{ij}$ $\forall (i,j) \in A$ for graph $G(N+S,A,B,T)$
>Perform forward or backward search on arcs with $u_{ij} > 0$ to find a path from S to sink t
>While there exists a path with positive flow from S to t
>Begin
>
>>Find a path P from S   to sink. Let $\delta = \min\{u_i: (i,j) \in P\}$;
>>Augment the flow on the path with δ ($u_{ij} = u_{ij} - \delta$ $\forall (i,j) \in P$);
>
>End While

End

The result of the algorithm is the set of paths from sources to the sink. For a dynamic flow network, the paths include the amount of flow and the time for evacuation. The traffic simulation model presented in Chap. 3 is used to simulate the resulting path chains. The simulation model simulates the flow along each arc. At the specified starting time, a flow leaves the source node and enters the first cell on the starting arc. At this time, the simulator assigns and tracks a path identification number to that flow. This identification number is a reference to path information that include flow amount, start time, and arcs sequence. The flow is then simulated through the cells on the arc until it reaches the end of the arc (the head node). At that time, and according to the assigned path number, the next arc to be traversed is determined. The simulation continues until all path chains reach safety.

5.3   Case Studies

Several case studies are developed that are based on parts of Houston and Galveston roads and highways [16]. Each case study needs the following parameters: network topology, roads distances, free flow speeds, capacities (maximum flow rate), jam densities, and the number of lanes. However, since the purpose of these case studies is to demonstrate the contribution of a new methodology and not for a verifiable

Table 5.1 Free flow speed and the number of lanes used in building the case studies

Road segment	Free speed (miles/h)	No. of lanes
Beltway 8	65	3
Major Interstate and US Highway (I45, I10, I610, US59)	60	3
State highways (TX6, TX225, TX288)	50	2
Local roads	45	2

Fig. 5.2 Flow rate–density diagram

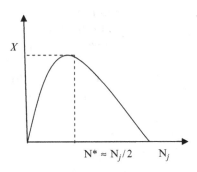

$$N^* \approx N_j/2 \qquad N_j$$

traffic model, no real data collection is involved. As an alternative, network parameters are determined as follows:

- *Network structure and distances*: The research relied on Microsoft Streets and Maps software to build network topology and measure distances.
- *Speed*: Free flow speed and the number of lanes are based on general public knowledge of Houston area. The following guidelines are used (Table 5.1):
- *Jam density*: Assuming an average vehicle length of 9 ft and a minimum headway (distance between vehicles on a jammed road) of 1 ft, then an average maximum density of 100 (veh./mile/lane) can exit on a road. Multiplying by the number of lanes, we can have an estimate for jam density.
- *Capacity*: According to traffic literature, roads capacities are normally estimated through real traffic data by observing the maximum flow rate for a given average speed. As an approximation, however, we can use the fundamental equation for macroscopic variables to estimate maximum flow rate. Recalling the fundamental diagram relating the flow rate with density in Chap. 3 (Fig. 5.2).

The maximum flow rate is approximated to be when the density is at half of the jam density. Then by using the fundamental equation, capacities can be calculated by the equation,

$$\text{Maximum flow rate} = \frac{\text{free speed} \times \text{jam density}}{2}. \qquad (5.1)$$

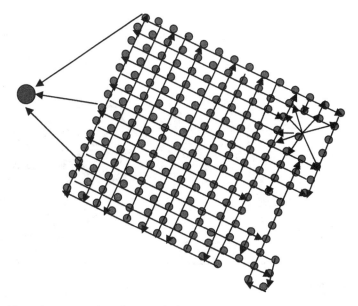

Fig. 5.3 Network representation of case study 1

5.3.1   Case Study 1

The first case study is taken from downtown Houston as shown below. The number of nodes is 158 and number of arcs is 291. The sink node is the largest size node in Fig. 5.3.

5.3.2   Case Study 2

Similarly, case 2 is taken from southwest side of Houston. The number of nodes is 45, and arcs 144. The sink node is at downtown Houston (Fig. 5.4).

5.3.3   Case Study 3

Case 3 is parts of the north side of Houston. The sink is on the northwest side of Houston. The number of nodes is 47 and arcs 163 (Fig. 5.5).

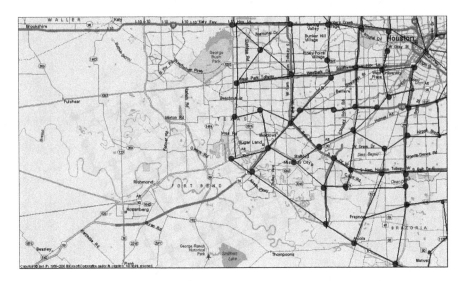

Fig. 5.4 Case study 2

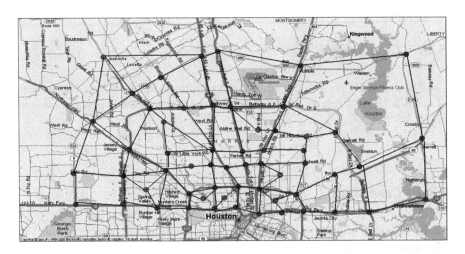

Fig. 5.5 Case study 3

5.3.4 Case Study 4

Case 4 is a comprehensive network covering most parts of Houston and Galveston. The sink node is also at the northwest side of Houston. The number of nodes is 181 and the number of arcs is 645 (Fig. 5.6).

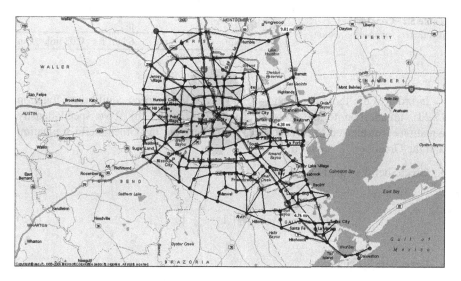

Fig. 5.6 Case study 4

5.4   Experimentation

The simulation model presented in Chap. 3 is repeated here. In this model, roads are divided into homogenous cells, and the state of the system is defined at each time step using the following state equations:

$$n_{ijk}(t+1) = n_{ijk}(t) + \frac{l_{ijk}}{\Delta t}[x_{ijk-1}(t) - -x_{ijk}(t)], \tag{5.2a}$$

$$v_{ijk}(t+1) = v_{ijk}(t) + \frac{\Delta t}{\tau}[V_{e}(n_{ijk}) - v_{ijk}(t)] + \frac{\Delta t}{l_{ijk}}v_{ijk}(t)[v_{ijk-1}(t) - v_{ijk}(t)] \\ -\frac{\mu\Delta t}{\tau l_{ijk}}\left[\frac{n_{ijk+1}(t) - n_{ijk}(t)}{n_{ijk}(t) + \kappa}\right], \tag{5.2b}$$

$$x_{ijk}(t) = v_{ijk}(t)n_{ijk}(t), \tag{5.2c}$$

$$V_{e}(n_{ijk}) = v_{f}\left(1 - \left(\frac{n_{ijk}}{N_{ijk}}\right)^{l}\right)^{m}, \quad l > 0, \quad m > 1, \tag{5.2d}$$

$$n_{ijk}(t+1) \le N_{ijk} \quad \text{or} \quad x_{ijk}(t) \le \frac{\Delta t}{l_{ijk+1}}[N_{iijk+1}(t) - n_{ijk+1}(t)] + x_{ijk+1}(t), \tag{5.2e}$$

$$x_{ijk}(t) \le u_{ijk}. \tag{5.2f}$$

Table 5.2 Constant
parameters for speed equation

κ	μ
40 veh./km = 72 veh./mile	6 km^2/h = 0.03 mile^2/min

In order to use these equations, we have to select a speed equilibrium function (5.2d), in addition to the constant parameters for (5.2b). Two speed–density equations are selected to be used in (5.2d). These are the GreenSheild's equation and the Greenberg's equation.

- GreenSheild's equation

$$v_{ijk} = v_{\mathrm{f}}\left(1 - \frac{n_{ijk}}{N_j}\right) \tag{5.3}$$

- Greenberg's equation

$$v_{ijk} = c \times \ln\left(\frac{n_{ijk}}{N_j}\right) \tag{5.4}$$

where c is constant, that is determined normally through historic data. Since no data collection is involved in this research, the following argument is used to determine "c." The simulation model is built on the assumption that vehicles travel approximately at free flow speed when densities are low. In Greensheild's equation, this assumption is satisfied. In Greenberg's equation we can find the value of the constant c that satisfies this assumption. Assuming that vehicles will travel at free flow speed if the current densities are at most 5% of jam density, then

$$v_{\mathrm{f}} = c \times \ln\left(\frac{0.05 \times N_j}{N_j}\right) \quad \text{or} \quad c = \frac{\ln(0.05)}{v_{\mathrm{f}}}. \tag{5.5}$$

5.4.1   Parameter κ, τ, and μ for Speed Equation (5.2b)

Recalling the discussion in Chap. 3, these parameters are estimated by calibrating a simulation model against real life date. In this research, we use the suggested values for these parameters. These parameters are given in Table 5.2.

We study the effect of parameter τ by choosing two levels for this parameter and solving each case study at each level.

5.4.2 Factors in Experimentation

1. Percentage of supply nodes to the total number of nodes: Two levels are selected: 20% and 30%.
2. The amount of supply at each node. Two levels are selected: 1,000 and 2,000. If the number of nodes exceeds 160, the levels are 500 and 1,000.
3. The value for parameter τ: Two levels are selected, 4 and 8.
4. Equilibrium speed equations: Two models are selected, Greensheild (GSHD) and Greenberg (GBRG).

The total number of these factor combinations is 16. Therefore, with four case studies, the total number of experiment runs is 64. Naturally, these problems could not be solved by hand, so a computer code is developed using C++ language. At each experiment run, three evacuation times are recorded:

- Optimum evacuation time using *constant travel time* quickest transshipment algorithm. This time is referred to as "*Optimum.*"
- The optimum constant travel time solution is decomposed in chain paths and run through the simulator for evaluation. The resulted evacuation time is recorded as "*Simulator.*"
- The proposed new methodology is used to solve the same run, and the resulted evacuation time is recorded as "*Router.*"

5.5 Experimentation Results

The results are summarized in Tables 5.3–5.10.

Table 5.3 Case 1 results with $\tau = 4$ (number of nodes is 158)

# of supply	31		47		31		47	
Total supply	31,000		47,000		62,000		94,000	
τ	4		4		4		4	
Optimum	65		96		127		190	
Speed model	GSHD	GBRG	GSHD	GBRG	GSHD	GBRG	GSHD	GBRG
Simulator	68	70	99	101	132	133	196	197
Router	67	68	97	98	129	130	191	192
Difference	−1	−2	−2	−3	−3	−3	−5	−5
% Reduction	1	3	2	3	2	2	3	3

Table 5.4 Case 1 results with $\tau = 8$

# of supply	31		47		31		47	
Total supply	31,000		47,000		62,000		94,000	
τ	8		8		8		8	
Optimum	65		96		127		190	
Speed model	GSHD	GBRG	GSHD	GBRG	GSHD	GBRG	GSHD	GBRG
Simulator	67	68	98	99	131	131	195	196
Router	66	66	97	97	128	129	191	191
Difference	−1	−2	−1	−2	−3	−2	−4	−5
% Reduction	1	3	1	2	2	2	2	3

Table 5.5 Case 2 results with $\tau = 4$ (number of nodes is 45)

# of supply	10		14		10		14	
Total supply	10,000		14,000		20,000		28,000	
τ	4		4		4		4	
Optimum	47		62		87		118	
Speed model	GSHD	GBRG	GSHD	GBRG	GSHD	GBRG	GSHD	GBRG
Simulator	58	69	84	119	108	163	178	244
Router	53	63	71	83	102	116	142	159
Difference	−5	−6	−13	−36	−6	−47	−36	−85
% Reduction	9	9	15	30	6	29	20	35

Table 5.6 Case 2 results with $\tau = 8$

# of supply	10		14		10		14	
Total supply	10,000		14,000		20,000		28,000	
τ	8		8		8		8	
Optimum	47		62		87		118	
Speed model	GSHD	GBRG	GSHD	GBRG	GSHD	GBRG	GSHD	GBRG
Simulator	57	60	72	75	98	101	129	132
Router	51	52	65	66	91	92	120	122
Difference	−6	−8	−7	−9	−7	−9	−9	−10
% Reduction	11	13	10	12	7	9	7	8

Table 5.7 Case 3 results with $\tau = 4$ (number of nodes is 47)

# of supply	10		14		10		14	
Total supply	10,000		14,000		20,000		28,000	
τ	4		4		4		4	
Optimum	35		39		57		68	
Speed model	GSHD	GBRG	GSHD	GBRG	GSHD	GBRG	GSHD	GBRG
Simulator	47	79	59	102	115	238	185	344
Router	45	64	54	76	73	120	92	168
Difference	−2	−15	−5	−26	−42	−118	−93	−176
% Reduction	4	19	8	25	37	50	50	51

Table 5.8 Case 3 results with $\tau = 8$

# of supply	10		14		10		14	
Total supply	10,000		14,000		20,000		28,000	
τ	8		8		8		8	
Optimum	35		39		57		68	
Speed model	GSHD	GBRG	GSHD	GBRG	GSHD	GBRG	GSHD	GBRG
Simulator	45	49	51	57	71	83	82	123
Router	43	46	47	48	63	64	75	75
Difference	−2	−3	−4	−9	−8	−19	−7	−48
% Reduction	4	6	8	16	11	23	9	39

Table 5.9 Case 4 results with $\tau = 4$ (number of nodes is 181)

# of supply	36		54		36		54	
Total supply	18,000		27,000		36,000		54,000	
τ	4		4		4		4	
Optimum	69		69		89		128	
Speed model	GSHD	GBRG	GSHD	GBRG	GSHD	GBRG	GSHD	GBRG
Simulator	107	471	190	546	312	781	543	1,093
Router	82	100	89	170	127	236	182	424
Difference	−25	−371	−101	−376	−185	−545	−361	−669
% Reduction	23	79	53	69	59	70	66	61

Table 5.10 Case 4 results with $\tau = 8$

# of supply	36		54		36		54	
Total supply	18,000		27,000		36,000		54,000	
τ	8		8		8		8	
Optimum	69		69		89		128	
Speed model	GSHD	GBRG	GSHD	GBRG	GSHD	GBRG	GSHD	GBRG
Simulator	82	88	87	97	127	176	190	290
Router	81	87	84	90	95	100	132	134
Difference	−1	−1	−3	−7	−32	−76	−58	−156
% Reduction	1	1	3	7	25	43	31	54

5.6 Analysis of Results

The result of the experiment shows improvement to the evacuation time in all 64 runs. The amount of improvement, however, varies from a 1% to 79% reduction in evacuation time. It is worth the effort here to investigate when the new router is expected to deliver more improvement. Let us consider the following example (Fig. 5.7 and Table 5.11).

Example 5.1 In this example, the numbers on the arcs represent travel time and capacity, respectively. The numbers above the nodes represent supply (or demand

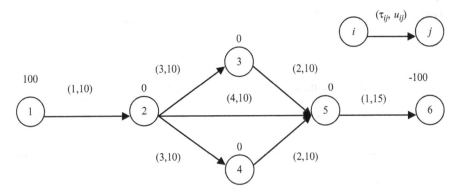

Fig. 5.7 Network representation for Example 7.1

Table 5.11 Network parameters for Example 7.1

Arc	Distance (mile)	Free speed (mile/min)	Jam density (veh./mile)	Capacity (veh./min)
(1,2)	1	1	20	10
(2,3)	3	1	20	10
(2,4)	3	1	20	10
(2,5)	4	1	20	10
(3,5)	2	1	20	10
(4,5)	2	1	20	10
(5,6)	1	1	30	15

if negative). If we solve the quickest transshipment problem for this example, it is not hard to see that the solution is by using path 1-2-5-6 ten times (with a flow of 10 each) and total evacuation time of 15 min. This example supports the general idea that to achieve minimum evacuation time, shortest paths should be used first as long as they are available (have positive residual capacity). In other words, path 1-2-5-6 is the shortest path in this example; and if we use any other path while the shortest path is available, the total evacuation time will be larger and not optimum anymore. This note is an introduction to the following analysis.

If we use the simulator to evaluate the optimum solution of Example 6.1 (using Greensheild's equation and $\tau = 4$), the resulting evacuation time is 18 min. Using the new router to solve Example 6.1, evacuation time is still 18 min with no improvement. Taking a closer look at the progress of building the solution for the new router reveals that path 1-2-5-6 maintains to be the shortest path from time step 0 until the end of iterations at time 18. Figure 5.8 (1)–(11) shows the first 11 iterations of the algorithm showing path 1-2-5-6 remains to be the shortest path (numbers inside the boxes are cell occupants, and numbers underneath the arcs are current travel times).

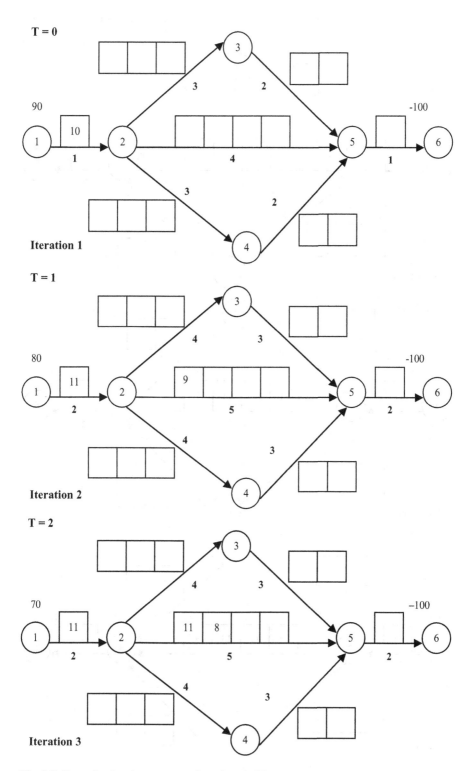

Fig. 5.8 Example of no improvement (*iterations 1–11*)

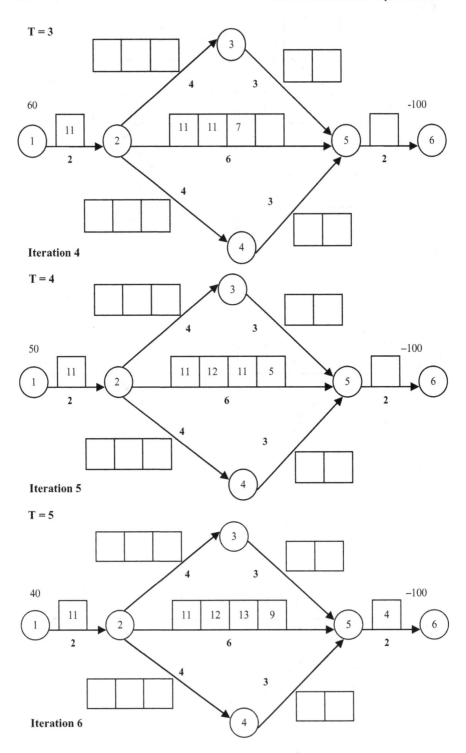

Fig. 5.8 (continued)

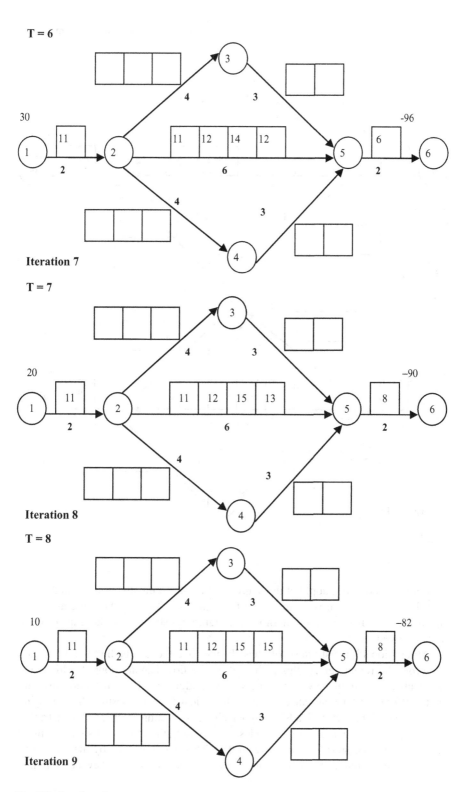

Fig. 5.8 (continued)

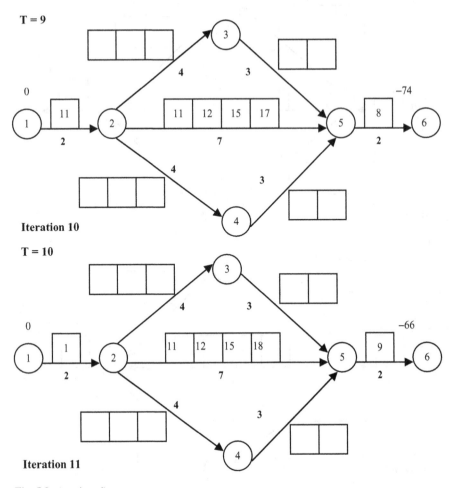

Fig. 5.8 (continued)

Let us assume the supply at the source is 140; the simulator solution is 24 min; and the router solution is 22 min. The first nine iterations are the same as shown above, starting from iteration number 10. Figure 5.9 shows how the improvement was achieved.

At iteration 13, the shortest path has changed and the new path is 1-2-3-5-6, and therefore the router selects this path to send the flow leaving node 2. This is the main advantage of the new router: the ability to adjust the routes if it finds any path that performs better than current routes. It is clear from the results that changing some experiment factors leads to higher chances of congestion or slower traffic. For example, adding more supply nodes or increasing the amount of supply is more likely to cause congestion and slower traffic. This is represented by the difference between the optimum calculated evacuation time and the simulated evacuation time

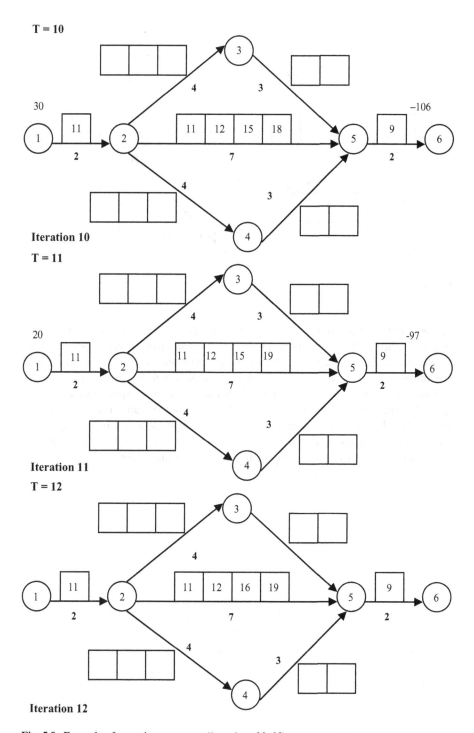

Fig. 5.9 Example of some improvement (*iterations 11–13*)

T = 13

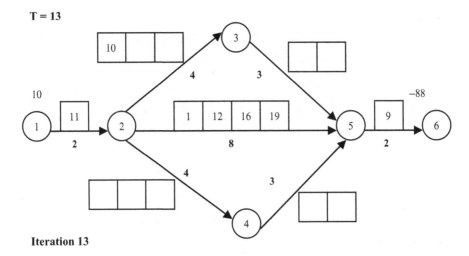

Iteration 13

Fig. 5.9 (continued)

for the same problem. In cases where supply increased the difference also tends to increase. Using Greenberg speed model as an equilibrium speed tends to cause slower traffic and more congestion than using the Greenshield equation. The speed parameter "τ" represents the rate at which average speed moves toward the equilibrium speed. A higher value for τ means that the average speed moves slower toward the equilibrium and thus affected slower by the amount of flow on the arc. In other words, the higher value of τ decreases the dependency of average speed on the flow on the arc and thus is less likely to cause congestion. The results supported this observation as we have slower traffic with $\tau = 4$ than when $\tau = 8$.

References

1. Blanchard BS (1998) Systems engineering management. Wiley Interscience, NJ
2. Systems Engineering Fundamentals (2001) Defense Acquisition University Press. Department of Defence, Systems Managements College, Fort Belvoir, Virginia
3. Reilly NB (1993) Successful systems engineering for engineers and managers. Kluwer Academic Publishers, Dordrecht
4. Boehm BW (1998) Spiral model of software development. In: Thayer RH, Dorfman M (eds) Tutorial software project management. IEEE Press, New York
5. Forsberg K, Mooz H, Cotterman H (2000) Visualizing project management: a model for business and technical success. Wiley, New York
6. Keane JF, Lutz RR, Myers SE, Coolahan JE (2000) An architecture for simulation based acquisition. Johns Hopkins APL Tech Dig 21:348–358
7. Matthews T (2003) A process review and appraisal of the systems engineering capability for the Florida Department of Transportation (FDOT), version 2, Technical Memorandum, February 20, 2003
8. Mitretek Systems, Inc. (2002) Building quality: intelligent transportation systems through systems engineering. Intelligent Transportation Systems Joint Program Office, US Department of Transportation
9. Technical Performance Measurement Handbook (1984) Department of Defense Management College, VA
10. U.S. DOT (2011) Research and innovative technology administration. Intelligent Transportation Systems Joint Program Office. http://www.its.dot.gov/index.htm
11. U.S. DOT (2007) Systems engineering for intelligent transportation systems. Federal Highway Administration & Federal Transit Administration
12. Cal. DOT. (2005) Systems engineering guidebook for ITS, ver. 1. California Department of Transportation
13. Goldblatt R (2004) Evacuation planning: a key part in emergency planning. In: Proceeding of 83rd annual meeting transportation research board. Washington, DC
14. Hamacher HW, Tjandra SA (2002) Mathematical modeling of evacuation problems: a state of the art. In: Sharma SD, Schrekenberg M (eds) Pedestrian and evacuation dynamics. Springer, Heidelberg
15. Greenshields BD (1935) A study of traffic capacity. In: Highway research board proceedings, vol 14, pp 448–477
16. Naser A (2008) An integrated methodology for dynamic routing optimization and road traffic simulation: application in evacuation planning. Dissertation, University of Houston

A. Naser and A.K. Kamrani, *Intelligent Transportation and Evacuation Planning:*
A Modeling-Based Approach, DOI 10.1007/978-1-4614-2143-6,
© Springer Science+Business Media, LLC 2012

17. Ahuja R, Magnanti T, Orlin J (1993) Network flows. Prentice Hall, NJ

18. Fahy RF (1995) EXIT89—an evacuation for high-rise buildings—recent enhancements and example applications. In: International conference of fire research and engineering. pp 332–337. September 10–15, 1995. Orlando, FL

19. Yamada T (1996) A network approach to a city emergency evacuation planning. Int J Syst Sci 27:931–936

20. Cova TJ, Johnson PJ (2003) A network flow model for lane-based evacuation routing. Transport Res 37:579–604

21. Ford LR, Fulkerson DR (1958) Constructing maximal dynamic flows from static flows. Oper Res 6:419–433

22. Minieka E (1973) Maximal lexicographic and dynamic network flows. Oper Res 21:517–527

23. Wilkinson WL (1971) An algorithm for universal maximal dynamic flows in a network. Oper Res 19:1602–1612

24. Hoppe B, Tardos E (1994) Polynomial time algorithms for some evacuation problems. In: Proceedings of 5th annual ACM-SIAM symposium on discrete algorithms. pp 433–441

25. Burkard RE, Dlaska K, Klinz B (1993) The quickest flow problem. ZOR Meth Model Oper Res 37:31–58

26. Chalmet LG, Francis RL, Saunders PB (1982) Network models for building evacuation. Manag Sci 28:86–105

27. Jarvis JJ, Ratliff HD (1982) Some equivalent objectives for dynamic network flow problems. Manag Sci 28:106–109

28. Hamacher HW, Tufekci S (1987) On the use of lexicographic min cost flows in evacuation modeling. Nav Res Logist 34:487–503

29. Wolfram S (1986) Theory and applications of cellular automata: including selected papers 1983–1986. World Scientific, NJ

30. Nagel K, Schreckenberg M (1992) A cellular automaton model for freeway traffic. J Phy I (France) 2:2221–2229

31. Klupfel H, Meyer-Konig T, Wahle J, Shreckenberg M (2000) Microscopic simulation of evacuation process on passenger ships. In: Fourth international conference on cellular automata for research and industry, October, Karlsruhe, Germany

32. Farahmand K (1997) Application of simulation modeling to emergency population evacuation. In: Proceedings of the 1997 winter simulation conference. pp 1181–1188. Washington, DC

33. Doheny J, Fraser J (1996) MOBEDIC—a decision modeling tool for emergency situations. Expert Syst Appl 10:17–27

34. Gartner NH, Messer CJ, Rathi A (1992) Traffic flow theory: a state of the art report—revised monograph on traffic theory. Federal Highway Administration. Washington, DC

35. Lighthill JJ, Witham GB (1955) On kinematic waves II: a theory of traffic flow on long crowded roads. Proc R Soc Lond A 229:317–345

36. Daganzo CF (1994) The cell transmission model: a simple dynamic representation of highway traffic. Transport Res B 28:269–287

37. Daganzo CF (1995) The cell transmission model, part II: network traffic. Transport Res B 29:79–93

38. Chien CC, Zhang Y, Stotsky A, Dharmasena SR, Ioannou P (1995) Macroscopic roadway traffic controller design. California PATH Research Report, UCB-ITS-PRR

39. Papageorgiou M (1989) Macroscopic modeling of traffic flow on the Boulevard Peripherique in Paris. Transport Res 23B:29–47

40. Corrine B (2000) Metacor—a macroscopic modeling tool for corridor application to the Stockholm test site. Research Report, RR-1998-0547, Center for Traffic Engineering and Traffic Simulation, Sweden

41. Heydecker BG, Addison JD (1997) Stochastic and deterministic formulations of dynamic traffic assignment. In: Proceedings of the 25th European Transport Forum. pp 107–120. London, England

42. Sussman J (2000) Introduction to transportation systems. Artech House.

43. Lu Q, George B, Shekhar S (2005) Capacity constrained routing algorithms for evacuation planning: a summary of results. University of Minnesota, pp 291–307
44. Haight FA (1958) Towards a unified theory of road traffic. Oper Res 6(6):813–826.
45. Greenberg H (1959) An analysis of traffic flow. Oper Res 7:78–85
46. Chen YL, Chin YH (1990) The quickest path problem. Comput Oper Res 17:153–161
47. Choi W, Francis RL, Hamacher HW, Tufekci S (1984) Network models of building evacuation problems with flow-dependent exist capacities. Proceeding of the 10th intranational Conference of Operations Research, Washington, DC, 1047–1059
48. Francis RL (1981) A uniformity principle for evacuation route allocation. J Res Natl Bur Stand 86(5)
49. Choi W, Francis RL, Hamacher HW, Tufekci S (1988) Modeling of building evacuation problems with side constraints. Eur J Oper Res 35:98–110
50. Fleischer L (1998) Efficient continuous-time dynamic network flow algorithms. Oper Res Lett 23:71–80
51. Kaufman DE, Nonis J, Smith RL (1998) Mixed integer linear programming model for dynamic route guidance. Transport Res B 32:431–440
52. Chiu YC (2004) Traffic scheduling simulation and assignment for area-wide evacuation. In: Proceedings of the IEEE intelligent transportation systems conference (ITSC), Washington, DC

Index

A. Naser and A.K. Kamrani, *Intelligent Transportation and Evacuation Planning:*   105
A Modeling-Based Approach, DOI 10.1007/978-1-4614-2143-6,
© Springer Science+Business Media, LLC 2012